Advances in Anatomy
Embryology and Cell Biology

Vol. 62

Editors

A. Brodal, Oslo   W. Hild, Galveston
J. van Limborgh, Amsterdam
R. Ortmann, Köln T. H. Schiebler, Würzburg
G. Töndury, Zürich   E. Wolff, Paris

Barbara Brown Gould

Organization of Afferents from the Brain Stem Nuclei to the Cerebellar Cortex in the Cat

With 10 Figures

Springer-Verlag
Berlin Heidelberg New York 1980

Dr. Barbara Brown Gould
Department of Medical Laboratory Sciences
P. O. Box 137
Quinnipiac College
Mt. Carmel Avenue
Hamden, Connecticut 06518
USA

The author's research, presented in this paper, was completed while the author was a graduate student at the Massachusetts Institute of Technology in the Department of Psychology and Brain Science, in partial fulfillment of the requirements for the degree of Doctor of Philosophy.

Library of Congress Cataloging in Publication Data. Gould, Barbara Brown, 1952 – The organization of afferents from the brain stem nuclei to the cerebellar cortex in the cat. (Advances in anatomy, embryology, and cell biology ; v. 62) Originally presented as the author's thesis, Massachusetts Institute of Technology. Bibliography: p. Includes index. 1. Afferent pathways. 2. Cerebellar cortex. 3. Brain stem. 4. Cats--Anatomy. 5. Mammals--Anatomy. I. Title. II. Series. [DNLM: 1. Brain stem. 2. Cerebellar cortex. 3. Neurons, Afferent. Wl AD433K v. 62 / WL310 G6960] QL801.E67 vol. 62 [QL938.A35] 574.4s [599.74'428]

ISBN-13: 978-3-540-09960-4   e-ISBN-13: 978-3-642-67614-7
DOI: 10.1007/978-3-642-67614-7

Composition: SatzStudio Pfeifer, Germering

2121/3321-543210

Contents

Acknowledgments

This work was supported by Grant BMS 75-18758 from the National Science Foundation, Grant NGR 22-009-826 from the National Aeronautics and Space Administration, Traineeship NIH-T01-GM01064 from the National Institute of General Medical Sciences, and by a Fellowship from the Massachusetts Institute of Technology Health Sciences Fund.

The author wishes to express her gratitude to Dr.A.M.Graybiel and the late Dr. H.-L. Teuber for their continuing advice and encouragement in the course of the author's graduate career, and to acknowledge the valuable comments of Drs. W.J.H. Nauta, E. Taber Pierce, E. Bizzi, J. Dekker, and P. Levitt. It is a pleasure to thank Mr. H. Hall for his skilled technical assistance.

Abbreviations

ß	nucleus ß of the inferior olivary complex
BC	brachium conjunctivum
BP	brachium pontis
C	nucleus cuneatus
CP	corpus pontobulbare
D	dentate nucleus
d.a.o.	dorsal accessory olive
d.c.	dorsal cap of the inferior olivary complex
d.m.c.	dorsomedial cell column
DV	descending vestibular nucleus
ECN	external cuneate nucleus
f	cell group f
F	fastigial nucleus
G	nucleus gracilis
I	interposed nucleus
IO	inferior olivary complex (subdivisions indicated in Fig. 10)
LC	locus coeruleus
LR	lateral reticular nucleus
m.a.o.	medial accessory olive
MLF	medial longitudinal fasciculus
MV	medial vestibular nucleus
N. dorsolat.	dorsolateral pontine nucleus
N. lat.	lateral pontine nucleus
N. med.	medial pontine nucleus
N. paramed.	paramedian pontine nucleus
N. ped.	peduncular pontine nucleus
Nrt	nucleus reticularis tegmenti pontis
NV	spinal trigeminal tract
NVII	facial nerve
N. ventr.	ventral pontine nucleus
P	griseum pontis (subdivision indicated in Fig. 4)
PH	nucleus praepositus hypoglossi
PM	paramedian reticular nucleus
p.o.	principal olive
RB	restiform body
Rm	nucleus raphe magnus
Rp	nucleus raphe pontis
Rpa	nucleus raphe pallidus
spv	nucleus supravestibularis
SV	superior vestibular nucleus
v.l.o.	ventro-lateral outgrowth of the inferior olivary complex
Vm	motor nucleus of the trigeminal nerve
Vsp	spinal trigeminal nucleus
x	cell group x
XII	hypoglossal nucleus

1 Introduction

The afferent connections of the cerebellar cortex of the cat have been extensively investigated by Alf Brodal and his collaborators using retrograde degeneration methods. These experiments (reviewed in Larsell and Jansen 1972) established that cerebellar cortical afferents arise from widespread areas of the brain stem and spinal cord. Brain stem nuclei shown to provide input to the cerebellar cortex included the pontine nuclei, the medial and descending vestibular nuclei, vestibular cell group x, the lateral reticular nucleus, the perihypoglossal nuclei, the paramedian reticular nucleus, the inferior olive, and the external cuneate nucleus. In addition, the red nucleus and certain of the raphe nuclei were thought to send fibers to the intracerebellar nuclei, but not to the cortex.

With the advent of the horseradish peroxidase (HRP) technique, new information on the distribution and organization of cerebellar cortical afferents has recently become available. Thus Gould and Graybiel (1976) demonstrated that afferents to the cat cerebellar cortex arise from a previously undescribed lateral tegmental cell group at the level of the isthmus and from the intracerebellar nuclei, as well as from the classic precerebellar nuclei. Moreover, these studies showed that fibers from the vestibular nuclei, previously thought to be distributed only to the flocculonodular lobe and uvula, reach widespread areas of the cerebellar cortex. Experiments by other investigators have established that the cerebellar cortex of the cat receives afferents from certain of the raphe nuclei (Shinnar et al. 1975; Taber Pierce et al. 1977), from the cuneate and gracile nuclei (Cheek et al. 1975; Rinvik and Walberg 1975), and from the central cervical nucleus (Matsushita and Ikeda 1975; Wiksten 1975). Reinvestigation of the projections to the cerebellar cortex from the inferior olive (Brodal A et al. 1975; Brodal A 1976; Hoddevik et al. 1976; Brodal A and Walberg 1977a, b; Hoddevik and Brodal A 1977; Kotchabhakdi et al. 1978), the pontine nuclei (Hoddevik 1975; Hoddevik et al. 1977; Brodal P and Walberg 1977; Hoddevik 1977; Brodal A and Hoddevik 1978; Hoddevik and Walberg 1979), the lateral reticular nucleus (Brodal P 1975; Dietrichs and Walberg 1979), and the vestibular nuclei (Kotchabhakdi and Walberg 1978b) using HRP techniques has shown, in addition, that the organization of cerebellopetal projections from these classic precerebellar nuclei is far more complex than previously described.

The goal of this paper is to review the data from recent studies concerning brain stem projections to the cerebellar cortex of the cat and to compare these results with some observations from the author's own investigations. Of particular interest in the present context are findings bearing on the following issues:

1. Which brain stem nuclei send fibers to the cerebellar cortex and where do their axons terminate?

2. By reviewing the literature on the known afferents to these brain stem nuclei what conclusions can be drawn as to the types of neural input which arrive at various regions of the cerebellar cortex by way of relays in the brain stem precerebellar nuclei?

3. Does the afferent anatomic organization of cerebellar cortical areas support or contradict current notions concerning the functional organization of the cerebellar cortex?

1

The author's own observations tend to support recent descriptions of afferents to particular regions of cerebellar cortex based on HRP studies. However, the present studies also provide new information about the origin and cortical distribution of some of the inputs from the brain stem nuclei. To avoid unnecessary repetition the author's findings will be discussed in detail only where they provide new information or differ from existing accounts.

2 Material and Methods

The material for the present study consisted of 37 young adult cats in which the enzyme tracer substance HRP (Sigma Type VI) was injected into various parts of the cerebellar cortex. In every case, suboccipital craniotomy was performed on cats deeply anesthetized with pentobarbital sodium (Nembutal), and the HRP solution was injected under direct visual guidance by means of a hand-held 1-μ 1 Hamilton syringe. In making the injections the syringe needle was usually inserted just beneath the pial-glial membrane, parallel to the surface of the cerebellum, for a distance of several millimeters. Total amounts ranging from 0.1 – 94.0 μl of HRP solution (the usual concentration was 50 %) were delivered in doses of 0.01 – 0.04 μl as the needle was slowly withdrawn. This technique proved rather successful in making injections which were confined to superficial cerebellar cortex.

Following survival times of 4 – 88 h animals were killed by transcardial perfusion with solutions of 0% – 4% paraformaldehyde and/or 0 – 3.5% glutaraldehyde in 0.1 M sodium phosphate buffer (pH 7.2 – 7.4) with 1% – 5% sucrose. The blocks were usually post-fixed further and washed overnight in phosphate buffer with 15% sucrose. The following day the blocks were cut in transverse sections at 50 μm on a freezing microtome. Then from one-half to all of the sections were reacted in a solution containing diaminobenzidine (DAB, free base, Sigma) and hydrogen peroxide according to the method of Graham and Karnovsky (1966).

Of the 37 cases with injections of HRP solution into cerebellar cortex, 19 had injection sites which were termed "superficial". In these cases the injection site, including its most diffuse borders visible under darkfield illumination, was restricted to the cerebellar cortex and medullary rays. There was no spread of HRP into the central medullary mass or the intracerebellar nuclei. It was important that injection sites be confined superficially in order to eliminate the possibility of uptake by fibers providing afferents to the deep cerebellar nuclei or by axons of deep nuclear neurons which might loop dorsally prior to leaving the cerebellum. In 17 cats, however, there was some spread of HRP into the central white matter and/or deep nuclei. The injection sites in these cases were termed "nonsuperficial". Although data from these cases alone could not provide reliable topographical information on cerebellar cortical afferents, they provided important supplementary data on the areal distribution of inputs to the cerebellum.

One case with a cortical injection served as a control. In this case (CHCB 26) a superficial cortical injection was attempted and the animal was killed 4 h after the operation to check the possibility that the injection sites visible at the standard survival times of 24 – 48 h might not reveal the full extent of HRP diffusion near the time of injection.

Five cases without injections in cerebellar cortex provided additional controls. In three cases HRP solution was injected hydraulically, under stereotaxic guidance, into the deep cerebellar nuclei or overlying white matter. In one case HRP was deposited in the left dorsal and ventral cochlear nuclei for comparison with the two cases with flocculus injections in which the injection site also involved the cochlear nuclei to some extent. Finally, one uninjected case was processed and reacted in order to check for the occurrence of endogenous peroxidase activity in the brain stem and cerebellum.

Details of the protocols for each case included in this report are summarized in Table 1.

Table 1. Summary of the details of the protocols for each case included in the present report

Case	Injection site	Amount of HRP[a]	Survival time	Fixative[b]
Superficial injection sites in cerebellar cortex				
CHRCB 13	Bilateral vermis IV, V	0.6 μl	24 h	3.5% glutaraldehyde
CHCB 12	Bilateral vermis, V, VI, VII	1.4 μl	45 h	1% paraformaldehyde 2.5% glutaraldehyde
CHCB 20	Bilateral vermis V, VI, VII	1.8 μl	48 h	2% paraformaldehyde 2% glutaraldehyde
CHCB 23	Bilateral vermis V, VI, VII	2.6 μl	24 h	2% paraformaldehyde 2% glutaraldehyde
CHRCB 6	Bilateral vermis V, VI, VII	2.2 μl	24 h	4% paraformaldehyde
CHRCB 18	Left intermediate vermis V, VI, VII	0.3 μl	26 h	3.5% glutaraldehyde
CHCB 27	Bilateral vermis IX	0.8 μl	29 h	2% paraformaldehyde 2% glutaraldehyde
CHRCB 8	Bilateral vermis VIII, IX	1.5 μl	21 h	3.5% glutaraldehyde
CHRCB 17	Bilateral vermis IX, X	0.1 μl	24 h	3.5% glutaraldehyde
CHRCB 10	Left paramedian lobule	1.7 μl	24 h	3.5% glutaraldehyde
CHCB 17	Left crus I	Not measured	42 h	2% paraformaldehyde 2% glutaraldehyde
CHCB 18	Left crus I and simplex	4.0 μl	42 h	2% paraformaldehyde 2% glutaraldehyde
CHCB 28	Left crus I and simplex	1.7 μl	25 h	2% paraformaldehyde 2% glutaraldehyde
CHRCB 7	Left crus I and simplex	1.0 μl	23 h	3.5% glutaraldehyde
CHRCB 12	Left crus II	1.2 μl	23 h	3.5% glutaraldehyde
CHCB 6	Left dorsal paraflocculus	0.2 μl	45 h	1% paraformaldehyde 2.5% glutaraldehyde
CHCB 25	Left dorsal paraflocculus	0.4 μl	25 h	2% paraformaldehyde 2% glutaraldehyde
CHRCB 11	Left ventral paraflocculus	0.2 μl	27 h	3.5% glutaraldehyde
CHRCB 20	Left flocculus and cochlear nuclei	0.2 μl	25 h	3.5% glutaraldehyde

Table 1 (continued)

Case	Injection site	Amount of HRP[a]	Survival time	Fixative[b]
Nonsuperficial injection sites in cerebellar cortex				
CHCB 5	Left half of the cerebellum	94 μl of 34% soln.	40 h	1% paraformaldehyde 2.5% glutaraldehyde
CHCB 7	Left half of the cerebellum exclusive of the flocculus and vermis I, II, X	7.0 μl	48 h	1.5% paraformaldehyde 2.5% glutaraldehyde
CHCB 11	Left half of the cerebellum plus the left inferior colliculus	5.2 μl	48 h	1.5% paraformaldehyde 2.5% glutaraldehyde
CHCB 10	Bilateral vermis and para-vermis III, IV, V, VI, VII, VIII, IX	1.6 μl	46 h	1.5% paraformaldehyde 2.5% glutaraldehyde
CHCB 24	Bilateral vermis II, III, IV, V, VI, VII, VIII, IX	4.6 μl	23 h	2% paraformaldehyde 2% glutaraldehyde
CHRCB 4	Bilateral vermis III, IV, V VI, VII, VIII	2.7 μl	44 h	4% paraformaldehyde
CHCB 14	Bilateral vermis II, III, IV	2.0 μl	43 h	2% paraformaldehyde 2% glutaraldehyde
CHCB 13	Bilateral vermis V, VI, VII	3.0 μl	31 h	4% paraformaldehyde
CHCB 15	Bilateral vermis V, VI, VII	2.9 μl	40 h	2% paraformaldehyde 2% glutaraldehyde
CHCB 19	Bilateral vermis V, VI, VII	3.0 μl of 66% soln.	41 h	2% paraformaldehyde 2% glutaraldehyde
CHCB 21	Bilateral vermis V, VI, VII	4.3 μl	88 h	2% paraformaldehyde 2% glutaraldehyde
CHCD 22	Bilateral vermis V, VI	3.2 μl	48 h	2% paraformaldehyde 2% glutaraldehyde
CHRCB 5	Bilateral vermis VIII, IX	1.5 μl	45 h	4% paraformaldehyde
CHCB 16	Bilateral vermis VIII, IX	1.4 μl	42 h	2% paraformaldehyde 2% glutaraldehyde
CHRCB 3	Left hemisphere exclusive of the paramedian lobule and flocculus	3.0 μl	43 h	4% paraformaldehyde
CHRCB 9	Left ventral paramedian lobule	0.2 μl	26 h	3.5% glutaraldehyde

Table 1 (continued)

Case	Injection site	Amount of HRP[a]	Survival time	Fixative[b]
CHRCB 19	Left paraflocculus, flocculus, and brachium pontis	0.2 μl	26 h	3.5% glutaraldehyde
Control cases				
CC-1	No HRP injected			2% paraformaldehyde 2% glutaraldehyde
CHCB 26	Bilateral vermis V, VI, VII	1.6 μl	4 h	2% paraformaldehyde 2% glutaraldehyde
CHRCB 16	Left cochlear nuclei	0.1 μl	25 h	3.5% glutaraldehyde
CHDN 1	Central medullary mass above the right dentate nucleus	0.02 μl of 66% soln.	23 h	1% paraformaldehyde 2.5% glutaraldehyde
CHDN 2	Left fastigial nucleus	0.02 μl of 58% soln.	40 h	1% paraformaldehyde 2.5% glutaraldehyde
CHDN 3	Left dentate and interposed nuclei	0.06 μl	23 h	3.5% glutaraldehyde

[a] The HRP was in 50% concentration (by weight) in 0.9% saline except where otherwise indicated.
[b] Fixative solutions were made up in 0.1 M sodium phosphate buffer (ph 7.2–7.4) with 1%–5% sucrose.

3 Observations from the Author's Investigations

3.1 General Comments

The findings in the present experiments will first be summarized briefly by using illustrative material from cases with large nonsuperficial HRP injections centered in either the cerebellar vermis, the hemisphere, or nearly the entire left half of the cerebellar cortex. Following this introduction, the findings after injections of more restricted regions of the cerebellar cortex will be considered. These results will be presented in the form of short descriptions of cell-labelling in the brain stem nuclei following injections in various parts of the cerebellar cortex. The cell groups projecting to the cerebellar cortex will be mentioned in roughly anatomic order, from rostral to caudal.

Several points should be borne in mind concerning the results of the present experiments. First, the injection sites in this series are generally fairly large, even if superficial, including several lobules in the vermis or large sectors of lobules in the hemisphere. Thus, while the appearance of HRP-positive neurons in a precerebellar nucleus indicates that these cells send axons to targets within the injected area, it is not possible to determine from the present evidence alone which part of the injection site constitutes the terminal field of the projection. Second, all of the HRP injection sites cen-

5

tered in the vermis are at least to some extent bilateral. Therefore, the present material cannot provide any evidence concerning the laterality of afferents to the vermis. Third, in cases of the superficial series, HRP injections have been confined primarily to the superficial lamellae of the cerebellar folia. While this minimizes the risk of obtaining false positives due to uptake by transit fibers, afferents to basal lamellae of the cerebellar cortex may be missed. Fourth, the HRP technique occasionally yields false negatives (Nauta et al. 1974). Therefore, absence of cell-labelling in a nucleus after HRP injections in cerebellar cortex does not constitute proof that the nucleus does not send fibers to the cerebellar cortex.

Larsell's (1953) nomenclature will be used in referring to parts of the cerebellum. The terminology adopted by Taber (1961) will be used in describing brain stem nuclei, except as otherwise indicated.

3.2 Results of Large Nonsuperficial Horseradish Peroxidase Injections

3.2.1 Left Half of the Cerebellar Cortex

In three cases with nonsuperfical injections of HRP covering large portions of the cerebellar cortex (vermis and hemisphere) on the left side, cell-labelling occurs in widespread areas of the brain stem and in the intracerebellar nuclei. Figure 1 illustrates the extent of the injection site and the distribution of retrogradely labelled neurons in one of these cases (CHCB 7). In all three cases HRP-positive neurons are found bilaterally in all parts of the *pontine nuclei* and *nucleus reticularis tegmenti pontis*, although in the pontine nuclei they are more numerous contralaterally. There is some tendency for pontine cell groups which are heavily labelled on one side to be unlabelled on the other side. At the caudal border of the pons, HRP-marked neurons are found bilaterally in the wedge-shaped *nucleus corporis pontobulbaris* (as delimited by Taber 1961). Near the caudal pole of the nucleus reticularis tegmenti pontis labelled neurons are found in the *nucleus raphe pontis*. Within this nucleus HRP-positive cells are found progressively more dorsally at more caudal levels. At the level of the rostral pole of the motor trigeminal nucleus labelled neurons encircle the medial longitudinal fasciculus, apparently comprising a caudal part of the nucleus annularis (Taber 1961). A few HRP-positive cells are located in the rostrodorsal part of the *nucleus raphe magnus*.

At this level HRP-marked neurons are also found in the *lateral tegmentum*. A conspicuous band of labelled neurons extends between the rostral part of the superior olivary complex and the motor trigeminal nucleus, bilaterally. Other HRP-positive neurons appear singly or in clusters among the emerging rootlets of the motor trigeminal and facial nerves, predominantly ipsilaterally. A few HRP-positive neurons are found in the *principal sensory trigeminal nucleus* bilaterally, with ipsilateral predominance.

Further caudally, in cases CHCB 5 and CHCB 11 (both not illustrated), HRP-marked neurons are found bilaterally in the *superior vestibular nucleus*. In all three cases labelled neurons also appear in the caudal halves of the *medial* and *descending vestibular nuclei*. Scattered HRP-positive neurons are found throughout the caudal parts of this latter nucleus, but the greatest concentration of labelled neurons is in *cell group f*. Smaller numbers of HRP-positive neurons are found in other nuclei of the vestibular complex including *cell group x*, *cell group y*, and the *supravestibular nucleus*. In cases CHCB 5 and CHCB 11 the *interstitial nucleus* of the vestibular nerve con-

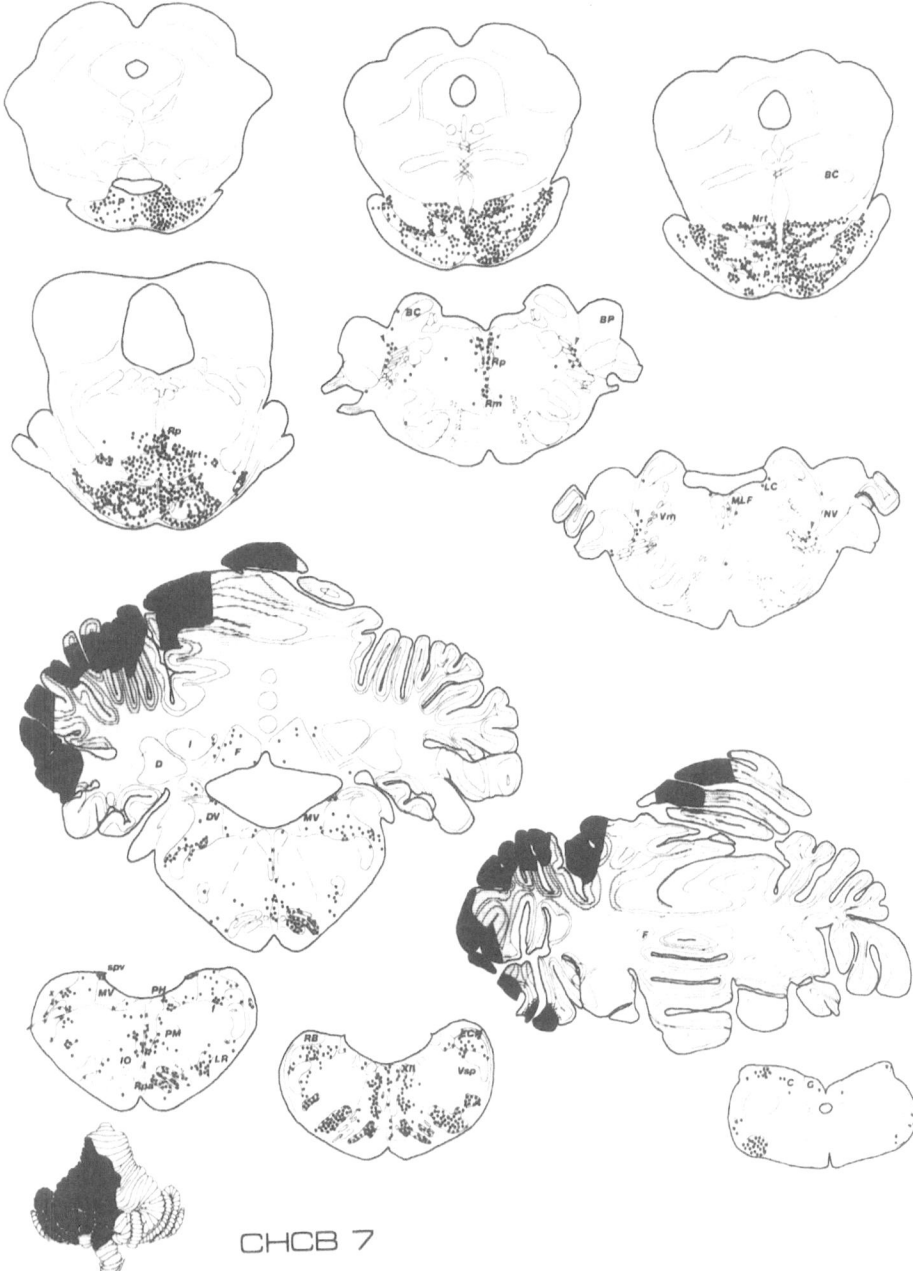

CHCB 7

Fig. 1. Chartings of selected transverse sections from case CHCB 7 with a large nonsuperficial injection site involving vermal and hemispheral cortex on the left side of the cerebellum. The approximate location of the injection site is indicated in *black* on a schematic drawing of the cerebellum (modified from Larsell, 1970) in this and the following figures. In transverse sections the darkest part of the injections site is indicated in *black, shaded areas* and the region bounded by a *dotted line* contain a still lighter HRP impregnation. *Black dots* indicate the locations of HRP-positive neurons and are spaced to represent the relative density of neurons within a given area

tains a few HRP-positive neurons. In all three cases labelled neurons appear bilaterally in the *nucleus praepositus hypoglossi*, primarily in its caudal half, although scattered HRP-positive cells can be found as far rostrally as the genu of the facial nerve. The other *perihypoglossal nuclei*, i.e. the *nucleus of Roller* and the *nucleus intercalatus*, contain somewhat fewer labelled cells. Contralaterally, all parts of the *inferior olivary nucleus* (except for the dorsomedial cell column in case CHCB 7) contain labelled neurons. (The olivary subdivisions are indicated on Fig. 10.) Some HRP-positive neurons also appear in the caudal medial accessory olive and nucleus ß ipsilaterally, reflecting the extension of the injection site across the midline. Labelled neurons are found in the *nuclei oralis* and *interpolaris* of the *spinal trigeminal complex*, with ipsilateral predominance. All parts of the *lateral reticular nucleus* contain HRP-positive neurons, bilaterally. Labelled neurons appear bilaterally in the dorsal, ventral, and accessory groups of the *paramedian reticular nucleus* and in the *nucleus interfasciculares hypoglossi*. Along the midline, HRP-positive neurons are located bilaterally in the *nucleus raphe obscurus* and *nucleus raphe pallidus*. At the most caudal levels cell-labelling occurs bilaterally in the *external cuneate nuclei*, the *main cuneate nuclei*, and the *gracile nuclei*. Within the nuclei cuneatus and gracilis, HRP-positive neurons are most numerous rostrally, but labelled neurons are scattered singly throughout the rostrocaudal extent of the nuclei. An occasional HRP-labelled cell is found within the *subnucleus magnocellularis* of the *nucleus caudalis* of the *spinal trigeminal complex* (not illustrated).

3.2.2 Vermis

The results in three cases with large HRP injections centered in the cerebellar vermis are similar in many respects to those described for large injections involving half of the cerebellum. Accordingly, the findings will be described more briefly, with emphasis on the differences between these results and those described above. Figure 2 shows the extent of the injection site and the distribution of brain stem cell-labelling in one of these cases, CHCB 24.

Following large vermal HRP injections in cases CHCB 10, CHCB 24, and CHRCB 4 labelled neurons are found bilaterally throughout the *pontine nuclei*. Horseradish peroxidase-positive neurons are located mostly in the pontine nuclei medianus, paramedianus, peduncularis, and dorsolateralis (boundaries as determined by Brodal A and Jansen, 1946; these boundaries are indicated on Fig. 4, and in the dorsal and lateral parts of the *nucleus reticularis tegmenti pontis*. A few labelled neurons of the nucleus reticularis tegmenti pontis are located as far laterally as the medial limit of the nucleus of the lateral lemniscus. Scattered HRP-positive cells are found in the *nucleus corporis pontobulbaris* and in the *lateral tegmentum* at the level of the isthmus. Labelled neurons are found in the *nucleus raphe pontis*, extending quite far dorsally. Some HRP-marked neurons which appear to be associated with this group extend quite far off the midline into the tegmentum and probably represent outlying neurons of the nucleus reticularis tegmenti pontis. A few labelled cells appear in the rostro-dorsal part of the *nucleus raphe magnus*. A single HRP-positive neuron is found within the *motor trigeminal nucleus* in case CHCB 24 (not illustrated).

The *medial* and *descending vestibular nuclei* contain many HRP-positive cells; *cell group f* is especially well-labelled. *Cell groups x* and *y* also contain HRP-marked neurons. Although not present in case CHRCB 4, in cases CHCB 10 and CHCB 24 (both not illustrated) the *superior vestibular nuclei*, the *interstitial nucleus* of the vestibular nerve, and the *supravestibular nucleus* contain a few HRP-positive cells.

All of the *perihypoglossal nuclei* contain labelled cells, as do all parts of the *lateral reticular nucleus* and *paramedian reticular nucleus*. A few HRP-positive neurons are found in the *nuclei oralis* and *interpolaris* of the *spinal trigeminal complex*. The *raphe nuclei pallidus* and *obscurus* contain a modest number of labelled cells. Cell-labelling in the *inferior olive* is confined primarily to the nucleus ß, the caudal medial accessory olive, and the dorsal accessory olive. In cases CHCB 10 and CHCB 24 (both not illustrated) the dorsomedial cell column also contains HRP-positive neurons. Labelled neurons are present in the *external cuneate nuclei*, primarily in the medial sector, and in smaller number in the *cuneate* and *gracile nuclei*.

3.2.3 Hemisphere (Excluding the Paramedian Lobule and Flocculus)

Figure 3 shows the extent of the injection site and the resultant brain stem cell-labelling in a case with a large HRP injection centered in the hemispheral cerebellar cortex (CHRCB 3). Labelled neurons are found throughout the *pontine nuclei*, in large numbers on the contralateral side, and in much smaller numbers on the side of the injection. Clusters of HRP-marked neurons are found contralaterally in the pontine nuclei paramedianus, ventralis, and lateralis. The medial half of the *nucleus reticularis tegmenti pontis* is filled with labelled cells, bilaterally. At the level of the isthmus there are a few HRP-positive neurons in the *corpus pontobulbare* (not illustrated) and in the *lateral tegmentum*. The latter are distributed predominantly ipsilaterally.

Cell-labelling in the *nucleus raphe pontis* extends dorsally almost to the floor of the fourth ventricle. At this most dorsal level, HRP-positive neurons encircle the medial longitudinal fasciculus. Just lateral to the raphe, many HRP-marked neurons are scattered in the tegmentum. These appear to be outlying cells of the *nucleus reticularis tegmenti pontis*, which extend quite far dorsally according to Taber (1961). The *medial* and *descending vestibular nuclei* contain HRP-positive cells, although they are not very numerous. Labelled neurons are found in the contralateral *inferior olive*, in the rostral medial accessory olive, and in the dorsal and ventral lamellae of the principal olive. A few HRP-positive neurons appear in the *perihypoglossal nuclei* and *paramedian reticular nucleus*, predominantly ipsilaterally. Very few labelled neurons can be found within the *lateral reticular nucleus* and the *external cuneate nucleus*.

3.3 Results of Restricted Cortical Horseradish Peroxidase Injections

3.3.1 Anterior Lobe

There are two cases with injection sites centered in the anterior lobe. In CHRCB 13 the superficial injection site is centered slightly to the left of the midline and involves the vermal and intermediate cortex of lobules IV and V bilaterally. The extent of the injection site and distribution of cell-labelling in the pons and other brain stem nuclei is this case are illustrated in Figs. 4 and 6. The injection site in CHCB 14 involves large sectors of lobules II-IV and encroaches on the central white matter and rostral part of the fastigial nuclei.

CHRCB 4

Fig. 2. Chartings of selected transverse sections from case CHRCB 4 with a large nonsuperficial injection site involving parts of the vermis bilaterally

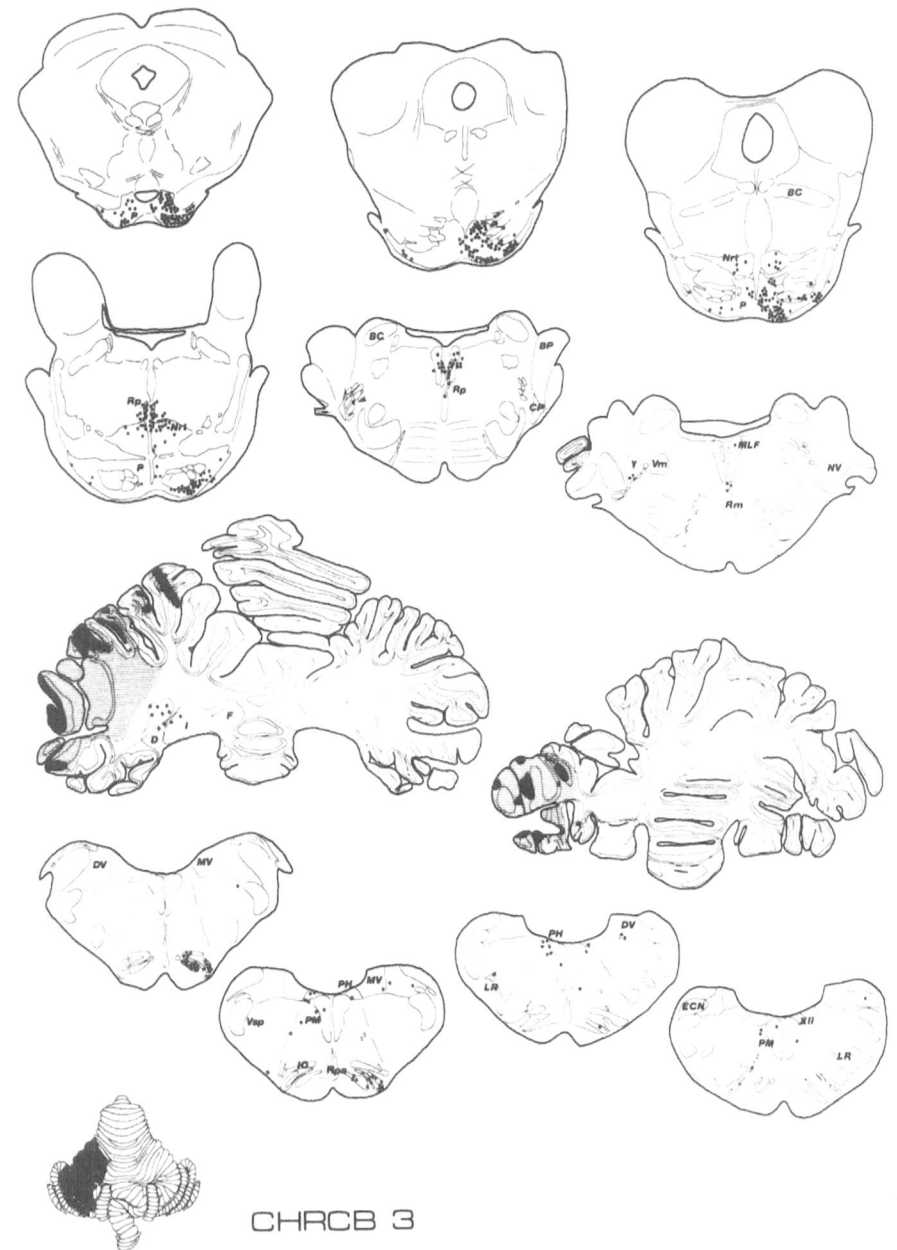

CHRCB 3

Fig. 3. Chartings of selected transverse sections from case CHRCB 3 with a large nonsuperficial injection site involving parts of the hemispheral cortex on the left side of the cerebellum

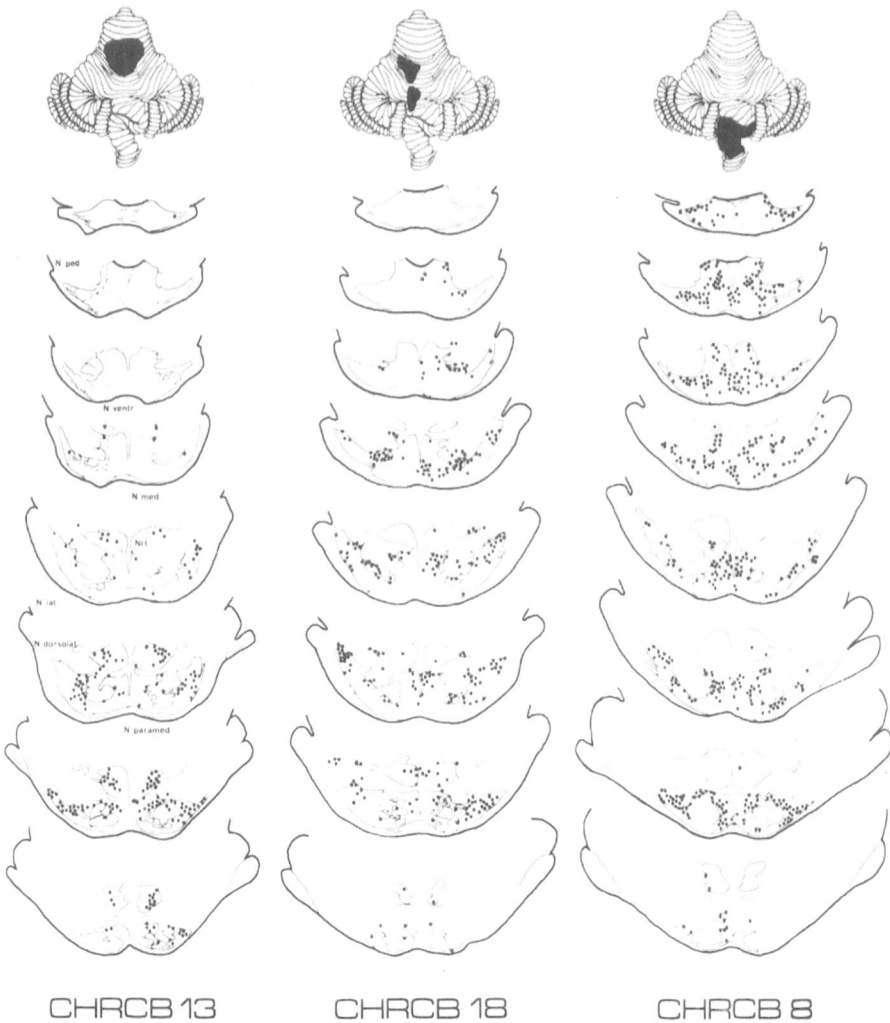

CHRCB 13　　　　**CHRCB 18**　　　　**CHRCB 8**

Fig. 4. Chartings of selected transverse sections through the pontine nuclei at approximately matched levels from each of three cases with injection sites centered in parts of the cerebellar vermis: CHRCB 13 (injection site in lobules IV-V), CHRCB 18 (injection site centered in the intermediate zone of lobules V-VII , and CHRCB 8 (injection site centered in lobules VIII- IX)

In both cases HRP-positive neurons in the *pontine nuclei* are found primarily in the caudal half of the pontine gray. The most conspicuous cell-labelling is in bands or clusters of neurons in the caudal parts of the nuclei peduncularis and lateralis and in a band occupying the middle third (from medial to lateral) of the *nucleus reticularis tegmenti pontis*. Smaller clusters of HRP-marked neurons are encountered in the nuclei paramedianus and dorsolateralis. There are a small number of labelled cells in the *nucleus corporis pontobulbaris*.

There are a relatively small number of HRP-positive neurons in the *raphe nuclei pontis*, *pallidus*, and *obscurus*. A few labelled cells are also found in the *lateral rhombencephalic tegmentum* in the area of the motor root of the trigeminal nerve

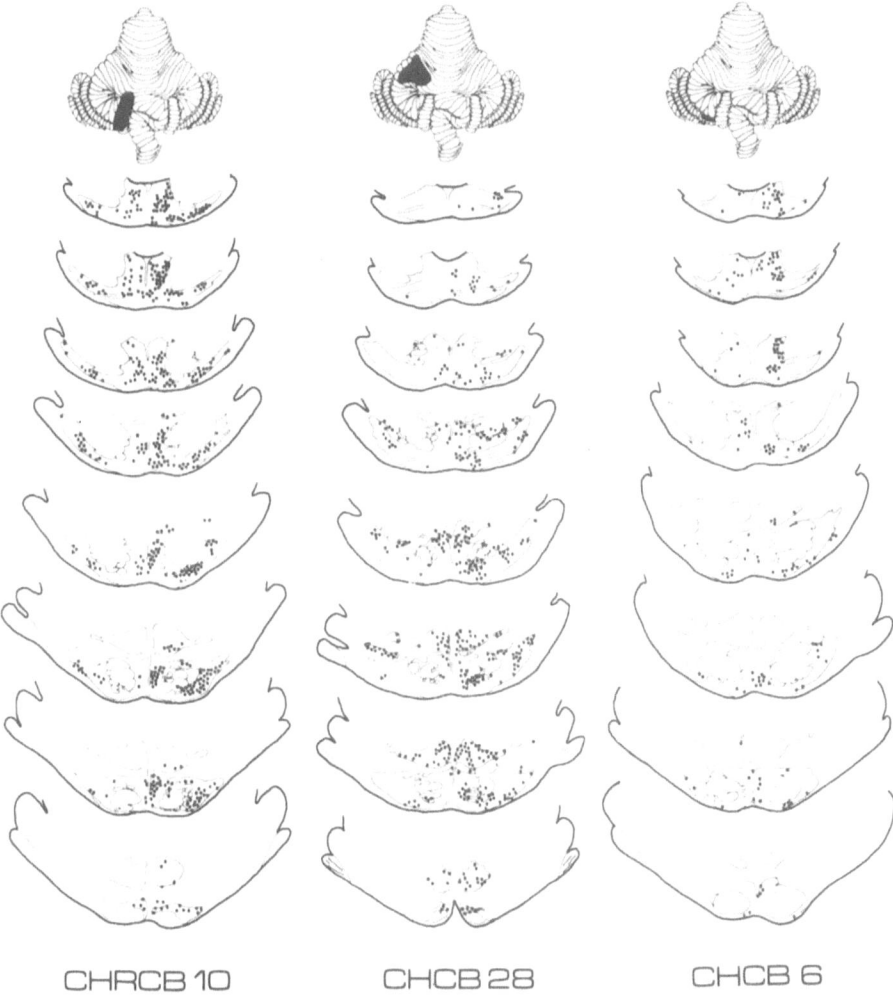

CHRCB 10   CHCB 28   CHCB 6

Fig. 5. Chartings of selected transverse sections through the pontine nuclei at approximately matched levels for each of three cases with injection sites centered in parts of the cerebellar hemisphere: CHRCB 10 (injection site centered in the paramedian lobule), CHCB 28 (injection site centered
in crus I), and CHCB 6 (injection site centered in the dorsal paraflocculus)

(arrowheads in Fig. 6). There is no apparent cell-labelling within the *trigeminal complex* except for a single HRP-positive neuron located in the *subnucleus magnocellularis*
of the *nucleus caudalis* of the *spinal trigeminal complex* in CHCB 14.

   Within the vestibular nuclear complex the *medial* and *descending vestibular nuclei*
contain a small number of labelled cells in case CHRCB 13. In the descending vestibular nucleus, most of the HRP-positive neurons are localized in *cell group f*. There is a
distinct band of labelled neurons within *cell group x*, and *cell group y* contains a few
labelled cells (not shown in Fig. 6). The distribution of HRP-positive cells within the
vestibular nuclear complex is basically the same in CHCB 14, except that the labelled
neurons are much more numerous, and there are a few HRP-positive cells in the *superior vestibular nucleus*.

Fig. 6. Chartings of selected transverse sections through the brain stem at approximately matched levels for two cases with injection sites centered in parts of the cerebellar vermis: CHRCB 13 (injection site centered in lobules IV-V) and CHRCB 18 (injection site centered in the intermediate zone of lobules V- VIII)

Fig. 7. Chartings of selected transverse sections through the brain stem at approximately matched levels for two cases with injection sites centered in parts of the cerebellar vermis: CHRCB 8 (injection site centered in lobules VIII-IX) and CHRCB 17 (injection site in the caudal part of lobule IX and lobule X)

15

Cell-labelling in the *perihypoglossal nuclei* is quite sparse in both cases, although labelled cells are contained within all three of the nuclei. In the *nucleus praepositus hypoglossi* HRP-positive neurons are most numerous caudally, while in the *nucleus intercalatus* cell-labelling is found predominantly in the rostral half.

The distribution of cell-labelling within the *inferior olivary complex* differs slightly on the two sides of the brain stem in CHRCB 13, probably reflecting the asymmetry of the injection site. On the right side HRP-positive neurons are found in the caudal medial accessory olive and in the caudal two-thirds of the dorsal accessory olive, predominantly in its lateral part. On the left side there are a few HRP-marked neurons in the caudal medial accessory olive. The overall distribution of olivary cell-labelling is similar in CHCB 14 except for some additional HRP-positive neurons in the caudal part of the dorsal lamella of the principal olive on one side.

In both cases many labelled cells are found in the magnocellular, parvicellular, and subtrigeminal portions of the *lateral reticular nucleus.* Many HRP-positive neurons are also clustered within the dorsal, ventral, and accessory cell groups of the *paramedian reticular nucleus.* Other HRP-marked cells are scattered singly around the outgoing fibers of the hypoglossal nerve in the *nucleus interfasciculares hypoglossi.* Labelled neurons are found throughout the rostrocaudal extent of the *external cuneate nucleus* in both cases. The *main cuneate* and *gracile nuclei* contain somewhat smaller numbers of HRP-positive neurons. Within these nuclei the labelled cells are most numerous rostrally.

3.3.2 Midvermal Lobules V-VII

There are ten cases with injection sites which straddle the border between the anterior and posterior lobes of the cerebellum, in the midvermal lobules V-VII. In five cases the injection sites are confined superficially, with no involvement of the central white matter or the deep nuclei. The injection sites in four of these cases are centered on the midline with varying degrees of spread laterally. The lateral spread of the injection site is smallest in CHCB 20 and CHCB 23, and greatest in CHCB 12. In CHRCB 18 (injection site also superficial) the injection site is centered to the left of the midline, with some spread of the injected HRP solution toward the midline (and to a very slight extent, across the midline) and laterally. The descriptions (below) of brain stem cell-labelling elicited by midvermal HRP injections will rely almost entirely on the results from these cases with superficial injection sites, with data obtained from the cases with nonsuperficial injection sites mentioned only where it provides additional information on brain stem-cerebellar projections. The extent of the injection site and distribution of labelled neurons in the pons and other brain stem nuclei in CHRCB 18 are illustrated in Figs. 4 and 6.

Of the cases with nonsuperficial injection sites, four have injections in lobules V-VII (CHCB 13, CHCB 15, CHCB 19, and CHCB 21) and one case has an injection in lobules V and VI only (CHCB 22). All of these injection sites are centered on the midline, with variable spread of the HRP solution laterally and to the central white matter and deep cerebellar nuclei. The lateral extension of the injection site is greatest in CHCB 15.

In the *pontine nuclei,* the neuronal cell groups projecting to lobules V-VII are located mostly in the caudal two-thirds of the pontine gray. There are only a few HRP-positive neurons in the rostralmost part of the pontine nuclei in cases of the superfi-

16

cial series, and these are scattered in the nuclei peduncularis, paramedianus, and medianus. In the caudal regions of the pons HRP-positive neurons appear in clusters in these nuclei and also in a band encompassing parts of the nuclei lateralis and dorsolateralis. In the *nucleus reticularis tegmenti pontis,* HRP-positive cells are located primarily in the dorsal half, with clusters of HRP-marked neurons extending as far laterally as the nucleus of the lateral lemniscus. There are a few labelled cells in the *nucleus corporis pontobulbaris* in all cases except CHCB 20.

A moderate number of retrogradely labelled neurons are encountered within the *raphe nuclei pontis, pallidus* and *obscurus.* There are also a few HRP-positive neurons in the rostrodorsal part of the *nucleus raphe magnus,* near its border with the caudoventral part of the nucleus raphe pontis.

A relatively large number of neurons in the *lateral rhombencephalic tegmentum* around the motor root of the trigeminal nerve are labelled by midvermal injections. In CHRCB 18 the HRP-positive neurons are found mostly on the left side, suggesting that this projection is predominantly ipsilateral. (The HRP-positive neurons are not present at the levels illustrated in Fig. 6, however.)

The *nuclei oralis* and *interpolaris* of the *spinal trigeminal complex* contain labelled neurons in cases CHCB 12, CHCB 20, CHCB 23, and CHRCB 6. There is a tendency for the HRP-positive neurons to be located at the lateral margins of the nuclei, near the spinal trigeminal tract. A very few labelled cells are found in the *subnucleus magnocellularis* of the *nucleus caudalis* of the *spinal trigeminal complex* in certain cases of both the superficial and nonsuperficial series (e.g. CHCB 20 and CHCB 15). In CHRCB 18 there is no evidence of cell-labelling within any of the trigeminal nuclei.

Several of the vestibular nuclear cell groups contain labelled neurons. The greatest number of HRP-positive neurons are found in *cell group f* of the *descending vestibular nucleus.* A smaller number are located in the caudal part of the descending vestibular nucleus outside of cell group f, in the *medial vestibular nucleus,* and in *cell group x.* Occasionally, HRP-positive neurons are found singly in *cell group y.* The small *nucleus supravestibularis* also contains labelled cells in one case, CHRCB 18 (not present at the levels drawn in Fig. 6).

There are labelled neurons throughout the rostrocaudal extent of the *perihypoglossal nuclei.* In the *nucleus praepositus hypoglossi* labelled neurons are most numerous in the caudal half. However, HRP-positive neurons are scattered singly in the rostral part of this nucleus, dorsal to the genu of the facial nerve. Cell-labelling in the *nucleus intercalatus* is found predominantly in its rostral half; only a few HRP-marked neurons are distributed in the caudal part of this nucleus.

The location of labelled neurons within the *inferior olivary nucleus* varies slightly in the different cases. Cell-labelling is restricted to the caudal medial accessory olive in case CHCB 20. In CHCB 23, the HRP-positive neurons are located predominantly in the caudal two-thirds of the medial accessory olive, but a smaller number of labelled cells are found in the medial part of the dorsal accessory olive. In CHRCB 18, labelled neurons are found in the caudal two-thirds of the medial accessory olive and in the dorsal accessory olive on the right side. There are also a few labelled cells in the dorsal lamella of the principal olive. On the left side the HRP-positive neurons appear in a cluster in the caudal medial accessory olive.

The magnocellular, parvicellular, and subtrigeminal portions of the *lateral reticular nucleus* contain many labelled neurons. A moderate number of HRP-positive cells are found in the dorsal. ventral, and accessory groups of the *paramedian reticular nucleus* and in the *nucleus interfasciculares of Jacobsohn.*

The *external cuneate nucleus* contains many labelled neurons throughout its rostrocaudal extent. A smaller number of HRP-positive cells are found in the *main cuneate* and *gracile nuclei,* except in CHCB 23, in which examination of the gracile nuclei did not reveal the presence of any neurons containing HRP reaction product. It is worthy of note that in the cases in which there is marked extension of the injection site into the intermediate and lateral zones of the midvermal lobules (e.g. CHRCB 18 and CHCB 15) the HRP-positive neurons are somewhat more plentiful in the cuneate and external cuneate nuclei than they are in the cases in which the injection site is restricted medially.

3.3.3 Posterior Vermal Lobules VIII-X

There are five cases with injection sites in posterior vermal lobules VIII-X. The injection sites are confined superficially in three of these cases: CHRCB 8, CHCB 27, and CHRCB 17. In CHRCB 8 the injection site involves most of lobules VIII-X bilaterally. The injection site in CHCB 27 is largely confined to lobule IX, also bilaterally, with slight spread to lobule VIII. The caudal part of the uvula and the adjoining part of the nodulus are included in the injection site in CHRCB 17. The extent of the injection site and the distribution of cell-labelling in the pons (CHRCB 8) and brain stem nuclei for cases CHRCB 8 and CHRCB 17 are illustrated in Figs. 4 and 7.

Of the two cases with nonsuperficial injections (CHRCB 5, CHCB 16), both have injection sites involving lobules VIII and IX bilaterally, with spread to the central white matter, caudal parts of the fastigial nucleus, and the medial edge of the paramedian lobule.

Clusters of labelled neurons are found throughout the rostrocaudal extent of the *pontine nuclei* in the cases with superficial injection sites involving lobules VIII and IX. In cases CHRCB 8 and CHCB 27 bands of HRP-positive neurons are found in parts of the nuclei peduncularis, paramedianus, medianus, lateralis, and dorsolateralis. There are very few labelled neurons in the *nucleus reticularis tegmenti pontis*, and these are found primarily in its medial half. In the superficial and nonsuperficial cases involving lobule VIII (CHRCB 8, CHRCB 5, CHCB 16) there are also a small number of HRP-positive cells in the *nucleus corporis pontobulbaris.* In contrast, in CHRCB 17, only a few labelled neurons can be found within the pontine nuclei proper, and there is no evidence of cell-labelling within the nucleus reticularis tegmenti pontis or corpus pontobulbare.

There are relatively few HRP-positive neurons within the *raphe nuclei* in the cases with superficial injection sites in posterior vermis. A scattering of labelled cells can be found within the *raphe nuclei pallidus* and *obscurus* in cases CHCB 27 and CHRCB 17 (not present at the levels illustrated in Fig. 7), and in CHRCB 8 there are a small number of retrogradely labelled neurons in the *nucleus raphe pontis.*

In all of these cases there are a small number of HRP-marked neurons in the *lateral tegmentum* at the level of the isthmus. These are mostly present in the dorsal area surrounding the motor trigeminal root fibers (not present at the levels shown in Fig. 7, however). Parts of the *principal sensory trigeminal nucleus* and the *nuclei oralis* and *interpolaris* of the *spinal trigeminal complex* also contain HRP-positive cells in the cases with superficial and nonsuperficial injection sites in lobules VIII and IX. However, no neurons containing reaction product can be found in any of the trigeminal nuclei in CHRCB 17.

In cases CHCB 27 and CHRCB 8 there are modest numbers of HRP-positive neurons in certain of the *vestibular nuclei*. In CHRCB 8 a small number of labelled cells are scattered in the caudal parts of the *medial* and *descending nuclei* and *cell group f*. *Cell group x* contains several distinct clusters of labelled neurons, and HRP-positive cells also appear in *cell group y* (not shown). Fewer HRP-marked neurons are present in the vestibular nuclei in CHCB 27 and these are localized in *cell groups x* and *y*.

In contrast, the vestibular nuclei contain many labelled neurons in case CHRCB 17. A moderate number of HRP-positive neurons are scattered in the *superior vestibularis nucleus*. There are a great many labelled cells in the *medial vestibular nucleus* throughout its rostrocaudal and dorsoventral extent. A band of HRP-positive neurons appears in *cell group x*. The *descending vestibular nucleus* contains many retrogradely labelled neurons, of which only a few are in *cell group f*. Many labelled cells are also found in *cell group y*, as shown in Fig. 1 of Gould (1979).

There is a rather modest amount of cell-labelling in the *perihypoglossal nuclei* in cases CHRCB 8, CHCB 27, and CHRCB 17. All three of the perihypoglossal nuclei contain labelled neurons, but there is a slight preponderance in the *nucleus praepositus*.

Within the *inferior olivary complex* the location of cell-labelling varies with the location of the injection site in the posterior vermal lobules. In CHRCB 8, HRP-positive neurons are found in three fairly well circumscribed cell groups: the dorsomedial cell column, the nucleus ß, and the lateral part of the caudal medial accessory olive. A few HRP-marked neurons are also scattered singly in the ventral lamella of the principal olive and the dorsal accessory olive. The dorsomedial cell column and nucleus ß contain the greatest concentration of HRP-positive cells in case CHCB 27. Scattered HRP-positive neurons also appear in the rostral and caudal parts of the medial accessory olive, and a few labelled cells are found in the ventral lamella of the principal olive. The HRP-positive neurons are distributed similarly in CHRCB 17, except that no labelled cells are present in the rostral medial accessory olive.

There are very few retrogradely labelled neurons in the *lateral reticular nucleus* in these cases. In CHRCB 8 and CHCB 27, HRP-positive neurons are located within the magnocellular and parvicellular parts of the nucleus. In case CHRCB 17 the small number of HRP-positive neurons are found only in the subtrigeminal part (not shown in Fig. 7).

In CHRCB 8 and CHRCB 17 there are a modest number of HRP-positive neurons contained within the *paramedian reticular nucleus*. In CHRCB 8 the labelled cells are in the dorsal and ventral groups of the nucleus; in CHRCB 17 they are in the rostral part of the accessory group. No HRP-marked neurons can be found in any part of the paramedian reticular nucleus or in the nucleus interfascicularis hypoglossi in CHCB 27.

Labelled neurons are found in the *main cuneate* and *external cuneate nuclei* in CHRCB 8 and in the cases with nonsuperficial injection sites in lobules VIII-IX. In CHCB 27 there are only a few retrogradely labelled cells in the external cuneate nuclei, however, and this may be attributable to slight diffusion of the injected HRP solution into lobule VIIIb. There is no evidence of cell-labelling within the external cuneate nuclei in CHRCB 17, or in the main cuneate in CHCB 27 and CHRCB 17. A few HRP-marked neurons are present in the *gracile nuclei* in CHRCB 8, but the gracile nuclei contain no labelled cells in CHCB 27 or CHRCB 17.

Fig. 8. Chartings of selected transverse sections through the brain stem at approximately matched levels for two cases with injection sites centered in parts of the cerebellar hemisphere: CHCB 28 (injection site centered in crus I) and CHRCB 12 (injection site centered in crus II)

Fig. 9. Chartings of selected transverse sections through the brain stem at approximately matched levels for two cases with injection sites centered in parts of the cerebellar hemisphere: CHRCB 10 (injection site centered in the paramedian lobule) and CHCB 6 (injection site centered in the dorsal paraflocculus)

3.3.4 Crus I and II and Lobulus Simplex

There are four cases with superficial injection sites centered in crus I. In CHCB 17 the injection site is centered in crus I with slight spread of the injected HRP solution to crus II, but no evidence of spread to lobulus simplex. In CHCB 18, CHCB 28, and CHRCB 7 the injection sites involve both crus I and lobulus simplex, with slight spread to crus II in all cases and to the lateral parts of lobule V in CHCB 18. The extent of the injection site and the distribution of cell-labelling in the pons and other brain stem nuclei in CHCB 28 are show in Figs. 5 and 8. There is one case with a superficial injection site centered in crus II (CHRCB 12). In this case the injection site involves virtually all of the superficial cortex of crus II and also extends slightly into the adjoining areas of crus I, the dorsal paraflocculus, and the dorsal part of the paramedian lobule. The extent of the injection site and the distribution of cell-labelling in the brain stem nuclei in CHRCB 12 are illustrated in Fig. 8.

In the cases with injection sites centered in crus I, HRP-positive neurons are found throughout the rostrocaudal extent of the *pontine nuclei* proper. There are clusters of labelled neurons in the nuclei paramedianus, ventralis, and lateralis, and in the caudal part of the nucleus peduncularis. In the *nucleus reticularis tegmenti pontis*, HRP-marked cells are present bilaterally, predominantly in the medial half of the nucleus. There is also a smaller cluster of labelled neurons in the lateral corner of the main body of the nucleus reticularis tegmenti. A modest number of labelled neurons are present in the *corpus pontobulbare*. In CHRCB 12 the distribution of cell-labelling in the *pontine nuclei, nucleus reticularis tegmenti pontis,* and *corpus pontobulbare* is similar to the distribution after crus I injections, except that there are fewer neurons in the pontine nuclei proper in the area ventral to the peduncle.[1]

Cell-labelling in the *raphe nuclei* is more extensive in the cases with superficial injection sites centered in crus I than in cases with injections placed anywhere else in the cerebellar cortex. A great number of HRP-positive neurons are present in the *nucleus raphe pontis* and, at the caudalmost levels of this nucleus, cell-labelling extends along and just off the midline raphe all the way to the floor of the fourth ventricle. A few of these dorsal HRP-marked neurons lie in and around the fibers of the medial longitudinal fasciculus, apparently comprising a caudal part of the nucleus annularis. A modest number of HRP-positive neurons are present in the rostrodorsal border zone of the *nucleus raphe magnus.* Farther caudally a moderate number of labelled cells are found in the *raphe nuclei pallidus* and *obscurus*. In contrast only a modest number of retrogradely labelled neurons are present in the *raphe nuclei pontis* and *pallidus* in CHRCB 12.

In CHRCB 12 numerous HRP-marked neurons appear, mostly ipsilaterally, in the tegmental region between the motor trigeminal nucleus and the rostral pole of the superior olivary complex (arrowheads in Fig. 8). However, only a few HRP-positive neurons are present, bilaterally, in the *lateral tegmentum* in the cases with injection sites centered in crus I (arrowhead in Fig. 8).

The maximum trigeminal cell-labelling in any of the cases of the superficial series is found in CHRCB 12. In this case a scattering of HRP-positive neurons appears

---

1 The term "peduncle" is used here to denote the caudal continuation of the cerebral peduncles. The fiber bundles passing through the pontine nuclei are usually referred to as the corticobulbar and corticospinal tracts. The term peduncle is employed in this discussion for convenience.

ipsilaterally in the *principal sensory trigeminal nucleus,* subnucleus ventralis (not shown in Fig. 8), and clusters of HRP-marked neurons are found in the ipsilateral *nuclei oralis* and *interpolaris* of the *spinal trigeminal complex.* In CHCB 17, CHCB 18, CHCB 28, and CHRCB 7, HRP-positive neurons are scattered singly within the *trigeminal nuclei,* predominantly in the ipsilateral *nucleus oralis* of the *spinal trigeminal complex.* Somewhat fewer labelled neurons appear in the *nucleus interpolaris.* In CHCB 28, several HRP-marked neurons are found in the ipsilateral *principal sensory trigeminal nucleus,* subnucleus ventralis (not shown in Fig. 8).

Only a small number of neurons in the *vestibular nuclei* are labelled by HRP injections centered in crus I or crus II. These are located predominantly in *cell group f,* bilaterally. There are also HRP-positive neurons scattered singly in the *medial* and *descending vestibular nuclei.*

There are a modest number of retrogradely labelled neurons in all subdivisions of the *perihypoglossal nuclei* in CHRCB 12 (labelled cells not present at levels shown in Fig. 8), and only a few labelled neurons are encountered within these nuclei in the cases with injection sites in crus I.

In the four cases with crus I injection sites, labelled *inferior olivary* neurons are concentrated in the dorsal and ventral lamellae of the principal olive beginning at the rostral pole, in the rostral medial accessory olive, and in the dorsal cap. A smaller number of HRP-positive cells are found in the medial part of the dorsal accessory olive in cases CHCB 18 and CHCB 28. There are a few labelled neurons in the ventrolateral outgrowth (not shown in Fig. 8) in cases CHCB 28 and CHRCB 7. In the case with a crus II injection site the cell-labelling is localized in the dorsal and ventral lamellae of the principal olive, the medial part of the dorsal accessory olive, and the middle levels of the medial accessory olive (not shown in Fig. 8).

Relatively few HRP-positive neurons are found in the *lateral reticular nucleus* in cases with injection sites in crus I or crus II. In the four cases with crus I injection sites cell-labelling is greatest in the magnocellular subdivision of the lateral reticular nucleus. The retrogradely labelled cells are found bilaterally, but they are much more numerous on the side of the injection site. In CHRCB 12 a small number of HRP-marked neurons are concentrated in the subtrigeminal portion of the ipsilateral nucleus. No labelled neurons can be found on the contralateral side.

Scattered HRP-positive neurons are found bilaterally in the dorsal, ventral, and accessory groups of the *nucleus reticularis paramedianus* and in the *nucleus interfasciculares hypoglossi* after injections in crus I. Labelled neurons are most numerous in the dorsal and ventral groups. In CHRCB 12 labelled neurons are found exclusively in the *nucleus interfasciculares,* almost entirely on the side ipsilateral to the injection site (not shown in Fig. 8).

A very small number of HRP-marked cells appear in the ipsilateral *external cuneate nucleus* in two of the four cases with injection sites centered in crus I (CHCB 18 and CHRCB 7). In CHCB 18 there are also a few labelled neurons in the *main cuneate nucleus,* ipsilaterally. These are not found in the other cases with crus I injection sites. No HRP-positive neurons are present in the gracile nuclei in the cases with crus I injections. The crus II injection in CHRCB 12 failed to elicit cell-labelling in either the external cuneate nucleus or the dorsal column nuclei.

3.3.5 Paramedian Lobule

There are two cases with injection sites centered in the paramedian lobule. In CHRCB 10 the injection site is confined superficially and involves most of the dorsal folia of the paramedian lobule, with slight spread of the injected HRP solution to the dorsally adjoining parts of crus I and lobulus simplex, and to a very minor extent, to the medial edge of lobule VII. The ventralmost folia of the paramedian lobule are not included in the injection site, however. The extent of the injection site and the distribution of cell-labelling in the pons and other brain stem nuclei in this case are illustrated in Figs. 5 and 9. In CHRCB 9 the injection site is in the ventral third of the paramedian lobule with clear extension into the central white matter and the interposed nuclei.

In CHRCB 10 clusters of labelled cells are present in the *pontine nuclei* paramedianus and lateralis, and to a more limited extent in the nucleus peduncularis. The great majority of the retrogradely labelled neurons are located on the side contralateral to the injection site. There are scattered HRP-positive cells in the ipsilateral dorsolateral pontine nucleus and in the medial part of the *nucleus reticularis tegmenti pontis* (predominantly contralaterally). There are a moderate number of labelled cells in the *corpus pontobulbare*.

The *raphe nuclei pallidus* and *obscurus* contain only a few HRP-positive neurons in CHRCB 10, and cell-labelling is absent in the other raphe nuclei. No retrogradely labelled neurons can be found in any of these midline cell groups in CHRCB 9.

In both cases there are a few labelled neurons in the *lateral tegmental area* around the motor root of the trigeminal nerve at the level of the isthmus. Further caudally in CHRCB 10 a small number of HRP-positive neurons are encountered ipsilaterally in the border zone between the *nuclei oralis* and *interpolaris* of the *spinal trigeminal complex*. Cell-labelling is absent in the trigeminal nuclei in CHRCB 9, however.

The nuclei of the *vestibular complex* contain a fair number of HRP-positive neurons in CHRCB 10. A band of HRP-marked neurons appears ipsilaterally in *cell group x*. A moderate number of labelled cells are present bilaterally in the *medial* and *descending nuclei*. Within the latter, *cell group f* contains the greatest number of HRP-marked neurons.

A moderate number of retrogradely labelled neurons are present bilaterally in all three of the *perihypoglossal nuclei* in CHRCB 10. Within the *nucleus praepositus hypoglossi* most of the labelled cells are located caudally; in the *nucleus intercalatus* HRP-positive neurons are predominantly in the rostral part of the nucleus.

In CHRCB 10 there is cell-labelling in the contralateral medial accessory *olive* at its middle levels and in the caudal parts of the dorsal and ventral lamellae of the principal olive. A small cluster of HRP-marked neurons is also found in the dorsal accessory olive at its middle levels.

There are many labelled neurons in the subtrigeminal, magnocellular and parvicellular subdivisions of the *lateral reticular nucleus* in CHRCB 10. The HRP-marked neurons are present bilaterally, but there is strong ipsilateral predominance. A moderate number of HRP-positive neurons are found bilaterally in the dorsal and accessory groups of the *paramedian reticular nucleus* in this case, and a smaller number are present in the *nucleus interfasciculares hypoglossi*.

In CHRCB 10, HRP-marked neurons are present throughout the rostrocaudal extent of the ipsilateral *external cuneate nucleus*, with a preponderance of HRP-positive neurons medially. Labelled neurons also appear medially in the contralateral external cuneate nucleus (not shown in Fig. 9) but they are few in number compared to the ipsilateral cell-labelling. There are also a fair number of retrogradely labelled neurons in the ipsilateral *main cuneate nucleus*, predominantly in the rostral portion. The gracile nuclei do not appear to contain labelled neurons.

3.3.6 Paraflocculus

There are three cases with superficial injection sites centered in the paraflocculus. In CHCB 6 the injection site is in the caudal part of the dorsal paraflocculus on the left side. In CHCB 25, the rostral part of the left dorsal paraflocculus has been injected. In CHRCB 11 the injection site is centered in the caudal part of the ventral parafloc-culus, with slight extension into the dorsal paraflocculus and the flocculus. The extent of the injection site and the distribution of cell-labelling within the pons and other brain stem nuclei in CHCB 6 are illustrated in Figs. 5 and 9.

Within the *pontine nuclei* proper HRP-positive neurons appear in clusters in the nuclei paramedianus, ventralis, and lateralis, and in parts of the nucleus peduncu-laris, predominantly contralateral to the injection site. A few labelled cells are present in the dorsolateral nucleus. Labelled cells are also scattered singly throughout the *nucleus reticularis tegmenti pontis*. There are no striking differences in the distribution of HRP-positive neurons within these nuclei in the three cases, despite the variation in location of the injection site within the paraflocculus. However, in CHRCB 11 only, there are a few retrogradely labelled neurons in the *corpus pontobulbare*, in addition to the cell-labelling in the pontine nuclei proper and the nucleus reticularis tegmenti.

There is no cell-labelling within the nuclei of the vestibular complex except for a band of HRP-positive neurons in the ipsilateral *cell group x* in CHRCB 11.

The *nucleus praepositus hypoglossi* and the *nucleus intercalatus* contain a few HRP-positive neurons in CHCB 6 and CHRCB 11, but no labelled neurons can be found within the perihypoglossal nuclei in CHCB 25.

In all three cases HRP-positive neurons are found in the *inferior olivary complex* at the ventral edge of the rostral medial accessory olive. In cases CHCB 25 and CHRCB 11, retrogradely labelled cells are also found in the ventrolateral part of the princi-pal olive, near the point where the dorsal and ventral lamellae merge. In CHCB 6 there is a conspicuous band of cell-labelling in the border region between the ventro-lateral outgrowth and the dorsal lamella of the principal olive.

No retrogradely labelled neurons can be detected within the raphe nuclei, the lateral tegmental cell group, the trigeminal nuclei, the lateral and paramedian reti-cular nuclei, the external cuneate nuclei, or the dorsal column nuclei in these three cases.

3.3.7 Flocculus

There are two cases with injection sites centered in the flocculus. In CHRCB 20 the injection site involves the flocculus and dorsal cochlear nucleus. There is slight in-volvement of the ventral paraflocculus. Some reaction product is evident among the fibers of the trigeminal nerve on the left side, indicating that the tip of the injection syringe may also have hit the trigeminal nerve. Although some extracerebellar struc-tures have been injected in this case, CHRCB 20 has been classified as belonging to the superficial series since, within the cerebellum, the injection site is confined to the cerebellar cortex and folial white matter. In CHRCB 19 the injection site is centered in the left flocculus, but the injected HRP solution has spread to the ventral and dorsal paraflocculi, the central white matter, and the brachium pontis. This case has been classified as having a nonsuperficial injection site.

In both cases there is a moderate amount of cell-labelling in the *pontine nuclei* and *nucleus reticularis tegmenti pontis*. The distribution of labelled cells is quite similar to that observed following HRP injections in the paraflocculus. In CHRCB 20 there are a moderate number of labelled neurons in the *corpus pontobulbare*. Since there is some involvement of the paraflocculus in both cases (and in CHRCB 19 the brachium pontis is also involved) it is not clear whether the pontine cell-labelling is due in any part to the HRP injection site in the flocculus.

The findings with regard to the *raphe nuclei* are inconsistent. In CHRCB 20 the *raphe nuclei dorsalis, pontis, magnus,* and *obscurus* contain moderate numbers of labelled neurons. In CHRCB 19, however, no retrogradely labelled cells can be found within the raphe complex.

The *lateral tegmental area* around the motor root of the trigeminal nerve also contains many HRP-positive neurons bilaterally (with ipsilateral predominance) in CHRCB 20, but no labelled cells are present in this area in CHRCB 19.

In CHRCB 20 there are a few HRP-positive neurons in the *principal sensory trigeminal nucleus* and the *mesencephalic trigeminal nucleus*, ipsilaterally, and in the *motor trigeminal nucleus* bilaterally. Because the injection site in this case apparently encroaches on the trigeminal nerve, it is questionable whether any of the observed trigeminal cell-labelling in this case is attributable to the floccular injection site. No retrogradely labelled neurons can be found in the trigeminal nerve nuclei in CHRCB 19.

In both cases there are a moderate number of HRP-positive neurons scattered in the *superior vestibular nuclei*, bilaterally. *Cell group y* contains a fairly large proportion of labelled cells on the side ipsilateral to the injection site (see Fig. 2 of Gould 1979). There are a small number of HRP-positive neurons in the *medial* and *descending vestibular nuclei*, but the latter are not in *cell group f*. In CHRCB 20 there are a few HRP-marked neurons in *cell group x* on both sides and in the *supravestibular nucleus* ipsilateral to the injection site.

In both cases many retrogradely labelled neurons are present bilaterally within all three *perihypoglossal nuclei*. As is the case with projections from the perihypoglossal nuclei to other parts of the cerebellar cortex, the neurons projecting to the flocculus tend to be located caudally within the *nucleus praepositus hypoglossi,* and rostrally in the *nucleus intercalatus*.

The HRP-positive neurons are rather sharply localized within the *inferior olivary complex*. In both cases the labelled cells are found almost entirely within the ventrolateral outgrowth and the dorsal cap (Fig. 10). In CHRCB 20 a few neurons containing reaction product are scattered in the rostral medial accessory olive and the dorsal and ventral lamellae of the principal olive. This scattered cell-labelling may be due to the extension of the injection site into the paraflocculus.

No labelled neurons can be found within the *lateral reticular nucleus* in either case. However, in both cases HRP-marked cells are present bilaterally within the *paramedian reticular nucleus*. A cluster of labelled cells is present in the rostral part

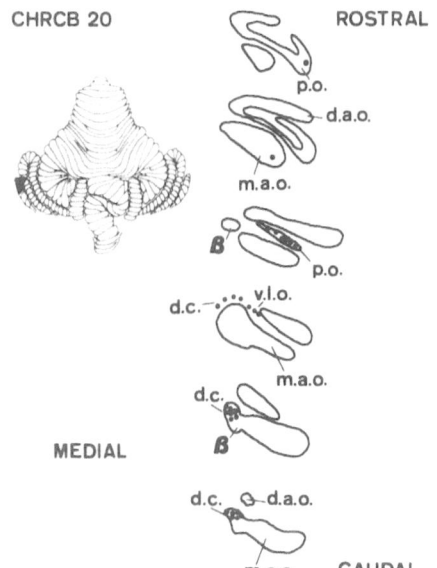

CHRCB 20   ROSTRAL

p.o.

d.a.o.

m.a.o.

B

p.o.

d.c.   v.l.o.

m.a.o.

d.c.

MEDIAL   B

d.c.   d.a.o.

m.a.o.   CAUDAL

Fig 10. Chartings of equally-spaced sections through the inferior olivary complex in case CHRCB 20 with an injection site centered in the flocculus

of the accessory group, and there are also a few HRP-positive neurons in the adjoining dorsal group.

In CHRCB 19 a few labelled neurons are present in the caudoventral part of the ipsilateral *external cuneate nucleus*. The external cuneate nuclei do not contain retrogradely labelled neurons in CHRCB 20, however. The dorsal column nuclei are negative in both cases with floccular injection sites.

Certain of the *cranial nerve motor nuclei* (in addition to the motor trigeminal nucleus, mentioned above) contain retrogradely labelled neurons. A modest number of HRP-positive neurons are present bilaterally in the *abducens nuclei* and *hypoglossal nuclei* in both cases. In CHRCB 20 the *oculomotor nucleus* contains a single labelled neuron.

The findings from the author's investigations are summarized in Table 2.

Table 2. Summary of the results of the present HRP studies demonstrating projections from brain stem nuclei to various parts of the cerebellar cortex

| | Injection site | | | | Hemisphere | | | | |
| | Vermis | | | | | | | | |
	IV–V	V–VII	VIII–IX	IX–X	Crus I and simplex	Crus II	Para-median lobule	Para-floc-culus	Floc-culus
Pontine nuclei	+++	+++	+++	?	+++	+++	+++	+++	?
Nucleus reticularis tegmenti pontis	++	+++	+		+++	+++	++	+	?
Corpus ponto-bulbare	++	+	+		+	+	++	+	?
Raphe nuclei	++	+++	+	+	+++	+	+		?
Lateral tegmental cell group	+	+++	+	+	+	+++	+		?
Trigeminal nuclei		++	+		+	+++	++		?
Vestibular nuclei	++	++	++	+++	?(+)	? (+)	? (+)		+++
Inferior olive	+++	+++	+++	+++	+++	+++	+++	+++	++
Perihypo-glossal nuclei	+	++	+	++	? (+)	? (+)	? (++)	+	++
Lateral reticular nucleus	+++	+++	++	+	++	+	+++		
Paramedian reticular nucleus	+++	+++	++	+	+	+	++		++
External cuneate nucleus	++	+++	++				+++		?
Cuneate nucleus	+	++	+				++		
Gracile nucleus	+	+	+						

4 Review of the Literature and Discussion of Brain Stem Afferents to the Cerebellar Cortex from Particular Precerebellar Nuclei

4.1 Pontine Nuclei, Nucleus Reticularis Tegmenti Pontis, and Corpus Pontobulbare

The pontocerebellar projection has been extensively investigated by Brodal and Jansen (1946) using retrograde degeneration methods. Although the results of more recent studies using the HRP technique show that the pontocerebellar projection is slightly more complex than originally described, certain organizing principles emerge from these pioneering studies which still appear to hold true. First, fibers passing to the cerebellar cortex from the pontine nuclei arise bilaterally, although the contralateral projection predominates. This is well illustrated by the author's own investigations (and practically all others) in cases in which the injection site is mostly restricted to the cerebellar cortex of one side (e.g., CHCB 7, Fig. 1; CHRCB 3, Fig. 3; CHRCB 10, CHCB 28, and CHCB 6, Fig. 5). Second, although neurons in the pontine areas which project ipsilaterally or contralaterally are not segregated into clearly defined cytoarchitectural areas, regions which project ipsilaterally are somewhat distinct from those projecting contralaterally. The original evidence is derived from Brodal and Jansen's findings in a case with a transection of most of the brachium pontis on one side; parts of the pontine gray which were heavily affected with retrograde changes on one side were less affected on the other. In the present series similar findings are obtained in cases with large injection sites involving most of the cerebellar cortex on one side: the locations of clusters of HRP-positive neurons on one side tend to correspond to the gaps in cell-labelling on the other side. Third, afferents to the vermal and hemispheral cerebellar cortex arise from largely complementary regions of the pontine gray. This notion is supported by the author's investigations in the cat (cf. Figs. 2 and 3) and by those of Eisenman in the rat (1976). Fourth, according to Brodal and Jansen (1946) practically all regions of the cerebellar cortex receive afferents from the pontine nuclei proper. Whether or not the flocculus and nodulus receive a projection could not be determined by retrograde degeneration techniques. All recent investigations using the HRP method support the notion that pontine fibers terminate over most of the cerebellar cortex. The nodulus does not appear to receive input from the pontine nuclei proper, however (present investigation as well as Voogd 1964; Hoddevik 1977), and whether the bulk of the flocculus received a projection is still uncertain (discussed below).

Retrograde degeneration experiments suggested that the projections to the vermis are derived from a midline strip of the pontine gray encompassing the medial part of the nucleus paramedianus, the nucleus medianus and the medial part of the nucleus ventralis, and from the nucleus dorsolateralis and the ventral half of the nucleus reticularis tegmenti pontis. Each lobule in the vermis appeared to be related to the total projection area in the pons. More recent experiments using the HRP method have established that the pontine regions sending fibers to different parts of the vermis are not identical. Furthermore, a large number of neurons in the peduncular nucleus and dorsal half of the nucleus reticularis tegmenti pontis project to parts of the vermis.

It is notable that the anterior lobe of the cerebellum receives pontine afferents from neurons located mostly in the caudal half of the pontine nuclei (Brodal P and Walberg 1977) and the middle third (from medial to lateral) of the nucleus reticularis tegmenti pontis (Hoddevik 1978), mainly contralaterally. The vermis of the anterior lobe is projected on by neurons located almost exclusively in the pontine gray lateral to the peduncle (mostly in the lateral and peduncular pontine nuclei) and in the central areas of the main part of the nucleus reticularis tegmenti pontis. Projections to the anterior lobules (I-III) arise more dorsally, and those to the posterior lobules (IV and V) arise more ventrally within the pontine nuclei proper. The projection from the nucleus reticularis tegmenti is relatively sparse, and there is no clear differentiation of areas which project to the various lobules. The intermediate-lateral parts of the anterior lobe receive input from two pontine regions, one located lateral to the peduncle and overlapping partly with the pontine region which projects to the vermis-proper of the anterior lobe, the other located medial to the peduncle in the paramedian and peduncular nuclei. As with the projection to the vermis proper, dorsal parts of the lateral area tend to project to the more anterior lobules, while ventral parts project to lobule V. The intermediate-lateral parts of the anterior lobe also receive a sparse projection from the main part of the nucleus reticularis tegmenti.

The cell groups in the pontine nuclei and nucleus reticularis tegmenti pontis which provide afferents to the vermis of the posterior lobe are somewhat differently situated than those providing afferents to the anterior lobe. According to Hoddevik et al. (1977) afferents to vermal lobules VI-VIII derive from neurons in four longitudinal columns which are situated largely within the dorsolateral, lateral, peduncular, and paramedian pontine nuclei (although the boundaries of the columns do not coincide precisely with the boundaries of these nuclei as defined by Brodal and Jansen 1946). Rather few of the neurons projecting to lobules VI-VIII are found in the rostralmost portions of these nuclei.

Hoddevik (1978) has reported that neurons projecting to these lobules are found throughout the rostrocaudal extent of the nucleus reticularis tegmenti pontis, in circumscribed cell groups situated dorsomedially and ventrolaterally within the main body of the nucleus and in a lateral part of the nucleus, known as the processus tegmentosus lateralis (Brodal A and Brodal P 1971). More neurons in the nucleus reticularis tegmenti are labelled by HRP injections centered in lobule VIIA than in other lobules. Injections in lobules VIIA and B and lobule VIIIA elicit cell-labelling in the processus tegmentosus lateralis, while those centered in lobules VI and VIIIB do not. The results of HRP experiments by Batini et al. (1978) with injection sites in lobules VI and VII also suggest that more neurons in the nucleus reticularis tegmenti pontis send fibers to lobule VII than to lobule VI.

There is one minor discrepancy between the findings of Batini et al. (1978) and those of Hoddevik (1978) which bears mentioning. According to Batini et al. (1978) HRP injections in lobule VII label a few neurons which extend out from the lateral part of the nucleus reticularis tegmenti pontis, and which are classified by them as belonging to the ventral nucleus of the lateral lemniscus. Examination of their Fig. 3 suggests that these may simply be outlying cells of the processus tegmentosus lateralis.

The results of anterograde degeneration experiments (Hoddevik 1978) also provide support for the notion that lobules VI and VIIA are major recipients of the efferents of the nucleus reticularis tegmenti pontis. Following lesions centered in this nucleus, the densest terminal degeneration is found within the granule cell layer of lobules VI and VIIA (and the flocculus).

Neurons which project to the uvula are localized in three longitudinally oriented columns which occupy the medial half of the peduncular nucleus and parts of the dorsolateral and lateral nuclei in the rostral two-thirds of the pons (Brodal A and Hoddevik 1978). A small number of neurons in the rostral part of the main body of the nucleus reticularis tegmenti pontis also send fibers to the uvula (Hoddevik 1978).

The results of the author's own investigations support Voogd's (1964) earlier suggestion that the nodulus does not receive a projection from the pontine nuclei proper in the cat. Similar results have been reported by Hoddevik (1977) in the rabbit. The nucleus reticularis tegmenti pontis apparently does provide a sparse projection to the nodulus in the cat, however (Hoddevik 1978). Axons passing to the nodulus arise from a rather circumscribed cell group in the dorsolateral part of the nucleus at caudal levels.

Retrograde degeneration experiments (Brodal A and Jansen 1964) suggested that the ansoparamedian lobule and lobulus simplex receive the bulk of their pontine input from the nuclei peduncularis and lateralis, the lateral parts of the nucleus paramedianus, the dorsolateral part of the ventral gray, and from parts of the nucleus reticularis tegmenti pontis. Crus I, crus II, the paramedian lobule, and the lobulus simplex all appeared to receive fibers from the entire pontine area related to the ansoparamedian lobule. More recent investigations using the HRP method largely support Brodal and Jansen's (1946) delineation of pontine cell groups projecting to the ansoparamedian lobule and lobulus simplex as a whole. However, these studies show that afferents to crus I, crus II, and the paramedian lobule arise to a large extent from separate neuronal cell groups within the pons.

The results of the author's own investigations show that neurons projecting to crus I and lobulus simplex are found bilaterally (with contralateral predominance) throughout the rostrocaudal extent of the pons in clusters located in the nuclei paramedianus, ventralis, and lateralis, and in the caudal part of the nucleus peduncularis. The cortex of crus I also receives a projection from neurons in the nucleus reticularis tegmenti pontis bilaterally. The majority of the neurons sending fibers to crus I and lobulus simplex are in the medial part of the nucleus, but some are also found in the lateral part of the main body of the nucleus reticularis tegmenti, and a very few are located in the processus tegmentosus lateralis, in agreement with Hoddevik (1978). Neurons projecting to crus II are found in a similar distribution within the pontine nuclei proper and nucleus reticularis tegmenti pontis as those projecting to crus I, except that relatively few neurons in the ventral pontine nucleus appear to project to crus II. A comparison of the distribution of cell clusters projecting to crus I and II suggests that the clusters are in adjacent, rather than overlapping, locations within the same general areas of the pontine gray. Hoddevik's (1978) studies suggest that mainly ventral areas of the nucleus reticularis tegmenti pontis project to crus I, while mainly dorsal areas project to crus II. This distinction was not obvious in the author's material however.

The paramedian lobule receives input from three longitudinal cell columns located medial to the peduncle (in the paramedian and peduncular nuclei), ventrolateral to the peduncle (in the peduncular and lateral nuclei and dorsolateral part of the ventral nucleus) and in the dorsolateral nucleus and adjoining parts of the lateral nucleus (Hoddevik 1975). Cells projecting to the paramedian lobule are found bilaterally, with contralateral preponderance, except for those in the dorsolateral nucleus which project mainly ipsilaterally. The projection from the medial cell column appears to be topographically organized. Dorsal parts of the column project to rostral folia, while ventral parts of the column project to caudal folia of the paramedian lobule. Neurons projecting to the paramedian lobule are found bilaterally in the ventral part of the main body of the nucleus reticularis tegmenti pontis (Hoddevik 1978).

The retrograde degeneration studies of Brodal and Jansen (1946) suggested that the pontine fibers to the paraflocculus of the cat derive predominantly from the contralateral lateral gray around the middle levels of the pons and from the paramedian gray rostrally. Recent HRP studies in the rabbit (Hoddevik 1977) show that pontine projections to the paraflocculus arise bilaterally (with contralateral predominance) from four pontine cell columns. Column "E" is located medial to the peduncle in the peduncular nucleus and adjoining part of the paramedian nucleus, rostrally. Column "F" is located ventral to the peduncle in adjoining portions of the peduncular, paramedian, ventral and lateral nuclei, mostly caudally. Column "G" is located in the dorsalmost part of the median nucleus and the dorsomedial part of the paramedian nucleus at mid-pontine levels. Column "H" is in a restricted region of the dorsolateral nucleus, caudally. According to Hoddevik (1977) neurons in all four columns project to the dorsal paraflocculus, while the ventral paraflocculus receives afferents only from columns E and F. It is unclear whether this is the case in the cat. In the author's own experiments an injection site centered in the ventral paraflocculus (CHRCB 11) elicited cell-labelling in all four columns. However, in this case there was very slight extension of the injection site into the dorsal paraflocculus, which may account for the result. In the cat, the paraflocculus also receives afferents from a fairly small number of neurons in the nucleus reticularis tegmenti pontis, mostly in the ventral part of the nucleus (Hoddevik 1978).

According to Hoddevik (1977), neurons in pontine cell column E project on the flocculus in the rabbit. However, Yamamoto (1979) has reported somewhat different results following flocculus injections in rabbits. Yamamoto's (1979) findings suggest that only folium p (classified by him as an extension of the ventral paraflocculus) receives afferents from the pontine nuclei proper; the main part of the flocculus does not. Furthermore, the location of the cell-labelling after Yamamoto's (1979) injections into folium p (his Fig. 2 d and e) seems to correspond more closely to Hoddevik's (1977) column F than to her column E.

In the present study in the cat, cell-labelling was present in Hoddevik's (1977) columns E, F, G, and H in both cases with injection sites centered in the flocculus. Since there was some involvement of the paraflocculus in the injection site in both cases (and in CHRCB 19 the brachium pontis is involved) it is not clear whether the observed cell-labelling is due in any part to the HRP injection site in the flocculus. However, it is worth noting that there is only minimal extension of the injection site into the ventral paraflocculus in CHRCB 20, and labelled neurons are still found within all four columns.

The flocculus receives afferents from a distinct cluster of neurons in the dorsomedial corner of the nucleus reticularis tegmenti pontis in the cat (Hoddevik 1978).

Except for the particular points raised above, the results of the author's own investigations on the organization of the pontocerebellar projection are in agreement with those of the Oslo school and do not contribute any additional information.

Comparison of the various pontine regions giving rise to projections to different cerebellar lobules indicates that a single, broadly defined cell column may contain neurons projecting to more than one cerebellar lobule. This raises a question as to whether single neurons in the pontine nuclei project to more than one lobule. Careful examination of chartings of distributions of retrogradely labelled neurons after HRP injections into various cerebellar lobules suggests that clusters of neurons projecting to different lobules lie adjacent to one another, rather than overlapping within the cell columns. However, a recent study using a retrograde fluorescent double labelling technique has shown that at least some pontine neurons send fibers to both vermal and hemispheral cerebellar lobules (Rosina et al. 1979).

Whether or not single pontine neurons generally project to more than one cerebellar lobule, there appears to be ample opportunity for convergence and divergence of inputs relayed by the pontocerebellar projection. Sensory and motor areas of the cerebral cortex, for example, project onto the pons in a divergent pattern, with axons terminating in two longitudinal cell columns, one medial and one lateral to the peduncle (Brodal P 1968a,b). A similar pattern of divergence is evident in the projection from the pontine nuclei to the cerebellar cortex. Particular pontine receiving areas do not send fibers to only one cerebellar cortical region, rather, different cell groups within each pontine receiving area send axons to several cortical regions. Furthermore, possibilities for convergence and integration of inputs from various modalities exist both in terms of the afferent organization of the pons and in terms of the pontocerebellar projection.

Afferents to the pontine nuclei proper have been described as arising from pyramidal cells in layer V of virtually the entire ipsilateral cerebral cortex (Kawamura and Chiba 1979), the superior colliculus, the inferior colliculus, the ventral lateral geniculate nucleus, the spinal cord, and the dentate and anterior interposed nuclei. Likewise, the nucleus reticularis tegmenti pontis is projected on by certain cerebral cortical areas (bilaterally); the superior colliculus; the lateral and superior vestibular nuclei; the fastigial, anterior interposed, and dentate nuclei; the central tegmental fields, and neurons in the vicinity of the oculomotor and trochlear nuclei. The available data suggest that no single input to the pontine nuclei is relayed exclusively to a particular region of the cerebellar cortex. There do, however, appear to be preferential areas of termination for projections from each pontine receiving area within the cerebellar cortex. While it is difficult to compare data from many different sources, it is of great interest to attempt to correlate information on the organization of the pontocerebellar projection with data on the termination of the various afferent systems within the pontine nuclei.

Within the pontine nuclei proper the dorsolateral nucleus is a major receiving area for teleceptive input. Afferents from the auditory cortex (Brodal P 1972a) superior colliculus (Altman and Carpenter 1961; Kawamura and Brodal A 1973; Graham 1977), and inferior colliculus (Kawamura 1975) terminate in largely overlapping areas within this nucleus. A large proportion of the neurons in the dorsolateral nucleus project to midvermal lobules VI-VIII (Hoddevik et al. 1977), which includes the visual and auditory receiving area of Snider and Stowell (1944). The most massive projection is to lobule VII (Kawamura 1975). Apparently, there is also a fairly substantial projection from this dorsolateral teleceptive area to crus I and II (author's investigations). Somewhat fewer neurons send axons to the cortex of the anterior lobe, and the uvula, paramedian lobule, and paraflocculus receive only a sparse projection. The superior colliculus also projects on a rather restricted dorsomedial sector of the nucleus reticularis tegmenti pontis, contralaterally. A smaller contingent of fibers terminates in the processus tegmentosus lateralis, ipsilaterally (Kawamura et al. 1974). These regions of the nucleus reticularis tegmenti send fibers mainly to lobules VI-VIII and crus I and II according to Hoddevik (1978).

Surprisingly, projections from the visual cortex do not terminate in the dorsolateral pontine nucleus in the cat along with the collicular inputs. Rather, visual cortical afferents terminate in bands within the rostral peduncular nucleus and in adjoining parts of the lateral, ventral, and paramedian nuclei (Brodal P 1972b,c; Glickstein et al. 1972 ; Baker et al 1976, Sanides et al. 1978). The largest proportion of neurons lying within the visual cortical receiving area appear to project to the uvula. Somewhat fewer neurons in this area project to the paramedian lobule, paraflocculus, and the cortex of the anterior lobe. The vermal visual area (lobules VI-VII) and crus I and II receive a relatively scanty projection from the pontine visual cortical receiving area. The visual cortex does not appear to send fibers to the nucleus reticularis tegmenti pontis.

Visual input also reaches the pontine nuclei by way of the ventral lateral geniculate nucleus, as has been reported by Graybiel (1974) and by Edwards et al. (1974). Both Graybiel (1974) and Ewards et al. (1974) indicate a terminal area for fibers from the ventral lateral geniculate in the rostral part of the nucleus paramedianus. Edwards et al. (1974) indicate a more dorsal termination than does Graybiel (1974), however. Neurons in the small pontine ventral lateral geniculate receiving area indicated by Graybiel (1974) appear to project mostly to the vermal visual area (lobules VI-VIII) and the uvula. Fewer neurons provide afferents to crus I.

Information concerned with body position and touch is apparently conveyed to the pontine nuclei and nucleus reticularis tegmenti by projections from the primary and secondary sensorimotor cortices and the spinal cord. Fibers from primary somatosensory cortex terminate in two separate areas of the pontine nuclei proper (one in the caudal part of the paramedian nucleus and one ventral and lateral to the peduncle, Brodal P 1968a). The major projection from these areas is to the cortex of the anterior lobe and the paramedian lobule, and a crude somatotopic representation appears to be preserved by these projections. Somewhat fewer neurons in the primary somatosensory receiving area project to the uvula. There is evidence also of a relatively sparse projection to lobules VI-VIII, crus I and II, and the paraflocculus. Axons from the primary somatosensory cortex also terminate in the ventral and medial parts of the nucleus reticularis tegmenti pontis (Brodal A and Brodal P 1971). The major projection from this area is to the anterior lobe and lobules VI-VIII, however, the uvula and crus I and II also receive input from a substantial proportion of the neurons in the ventromedial part of the nucleus reticularis tegmenti. A few neurons in the primary somatosensory projection field send axons to the paramedian lobule and paraflocculus.

Fibers from the secondary somatosensory cortex terminate in three longitudinally oriented columns in the pontine nuclei proper (Brodal P. 1968b), located medially, ventrolaterally, and dorsolaterally within the pontine gray. Only in the medial column does there appear to be any great amount of overlap with the projection area of primary somatosensory cortex. There appear to be substantial projections from the secondary somatosensory receiving area to the anterior lobe, the uvula, the paramedian lobule and the paraflocculus. (The projection to the paramedian lobule is somatopically organized.) Fewer cells in this region send axons to lobules VI-VIII and crus I and II. Secondary somatosensory fibers also terminate in the ventrolateral part of the nucleus reticularis tegmenti (including the processus tegmentosus lateralis), overlapping considerably with the terminal fields of fibers arising from other cortical areas (Brodal A and Brodal P 1971). This region of the nucleus reticularis tegmenti seems to project most heavily to lobules VI-VIII. The uvula and crus I and II receive a lesser projection, while only a relatively small number of cells provide fibers to the paramedian lobule and paraflocculus.

The motor cortex of the cat projects on two longitudinally oriented columns in the caudal half of the pontine nuclei (Brodal P 1968a). One column is located medial to the peduncle, in the peduncular nucleus and adjoining parts of the paramedian nucleus. The other column is located lateral to the peduncle, in the peduncular and lateral nuclei. The largest proportion of neurons in the pontine motor cortical receiving area project to the anterior lobe; somewhat fewer send axons to the paramedian lobule. There appear to be relatively sparse connections with lobules VI–VIII, crus I and II, and the paraflocculus. Motor cortical fibers also terminate in the ventral part of the main body of the nucleus reticularis tegmenti, overlapping almost entirely with terminal field of primary somatosensory axons (Brodal A and Brodal P 1971). The main projections from this part of the nucleus reticularis tegmenti are to the anterior lobe and lobules VI –VIII, with the uvula and crus I and II receiving a somewhat lesser projection. Scattered neurons in this area provide afferents to the paramedian lobule and paraflocculus.

The projection of the proreate gyrus onto the pontine nuclei and nucleus reticularis tegmenti coincides approximately with that of the motor cortex (Brodal P 1971a; Brodal A and Brodal P 1971). Therefore, the input from the proreate gyrus would appear to be relayed to the same areas of cerebellar cortex as the motor cortical input and will not be described separately.

The anterior orbital gyrus projects onto approximately the same regions of the pontine nuclei and nucleus reticularis tegmenti pontis as the secondary somatosensory cortex (Brodal P 1971b). Input from anterior orbital cortex can be expected, therefore, to reach the same cerebellar cortical regions as the input from secondary somatosensory cortex which is relayed by the pontine nuclei. The posterior orbital gyrus projects, in addition, on a longitudinally oriented column at the ventromedial aspect of the peduncle, where it does not overlap with the areas receiving input from the secondary somatosensory cortex. The largest number of neurons in this ventromedial column appear to send fibers to crus I and the paraflocculus.

Axons from the parietal cortex terminate in a fairly large band of the pontine gray, encompassing most of the outer circumference of the peduncular nucleus and adjoining parts of the paramedian, ventral and lateral nuclei (Mizuno et al. 1973). Many neurons in this region project to lobules VI–VIII, the paramedian lobule, crus I, and the paraflocculus. Somewhat fewer send axons to the uvula and crus II, and the intermediate-lateral zone of the anterior lobe receives input from a small number of cells in this area.

A recently published study by Ruegg et al. (1978) indicates that afferents from the spinal cord terminate in the caudalmost regions of the dorsolateral pontine nucleus in the cat, ipsilaterally. The area receiving spinal afferents in their Fig. 1 appears to correspond most closely to the area of the dorsolateral nucleus which projects to the anterior lobe and lobules VI -VIII. Fewer neurons in this area send axons to the uvula, paramedian lobule, and crus I and II.

The cerebellum itself provides input to parts of the pontine nuclei and nucleus reticularis tegmenti pontis. The dentate and anterior interposed nuclei send fibers to the pontine nuclei and nucleus reticularis tegmenti pontis by way of the descending limb of the brachium conjunctivum. These axons terminate in three longitudinal columns in the pontine nuclei: in the caudal part of the dorsolateral nucleus, in the caudal part of the peduncular nucleus (dorsally) and in the rostral part of the para-median nucleus (Brodal A et al. 1972),. Neurons in these areas project mainly to crus I and II, and to a lesser extent on the anterior lobe, posterior vermis, and paramedian lobule. Axons from the dentate and anterior interposed nuclei also terminate in the central region of the main body of the nucleus reticlaris tegmenti pontis (Brodal A and Szikla1972, Brodal A et al. 1972). This region provides afferents mainly to the anterior lobe and crus I and II, and to a lesser extent to lobules VI–VIII, the uvula, and the paramedian lobule.

There is no evidence for a projection from the fastigial nucleus to the pontine nuclei proper. However, fastigial axons pass to the dorsal part of the nucleus reticu-laris tegmenti pontis (Walberg et al. 1962). Neurons in this region project mainly to lobules VI-VIII and crus I and II and to a lesser extent to the paramedian lobule, flocculus and nodulus.

The superior and lateral vestibular nuclei also send fibers to the nucleus reticu-laris tegmenti pontis. Axons of both vestibular nuclei terminate in the caudoventral part, mainly contralaterally (Ladpli and Brodal A 1968). This sector of the nucleus reticularis tegmenti pontis provides afferents mostly to the anterior lobe and lobules VI-VIII, and to a lesser extent to the uvula, paramedian lobule, and crus I and II.

Neurons in the central tegmental fields also send axons to the caudoventral part of the main body of the nucleus reticularis tegmenti (Berman 1977). The terminal field of this projection appears to overlap to a great extent with that of the vestibular nuclei. Neurons in the region which receives input from the central tegmental fields project, in turn, mainly to the anterior lobe and lobules VI-VIII, and to a lesser extent to the uvula, paramedian lobule, and crus I and II.

Neurons in the region of the oculomotor and trochlear nucleiproject to a very circumscribed area in the dorsomedial part of the caudal nucleus reticularis tegmenti, according to a recent study by Graybiel (1977). It is of great interest that the terminal field for this projection corresponds rather closely to the very circumscribed regions of the nucleus reticularis tegmenti which have been shown to give rise to fibers passing to the flocculus and nodulus (Hoddevik 1978). Some neurons in this area also project to lobules VI–VIII.

It is not yet clear whether single neurons within the pontine gray receive synaptic inputs from more than one source, or whether single pontine neurons generally pro-ject to more than one cerebellar lobule. However, the considerable overlap in the ter-minal fields of various afferents in the pontine nuclei and the apparent overlap of pro-jection fields in the cerebellar cortex from various pontine receiving areas suggest that considerable integration of proprioceptive and teleceptive information could be achiev-ed by way of the organizational properties of the pontocerebellar projetion. This could be important, for example, in directing movement toward moving targets using auditory or visual cues. In this connection, it is of interest that recent physiologic stu-dies have shown that the auditory and visual information which is made available to the cerebellum is concerned not with identification, but with the position, direction, and velocity of movement of visual and auditory targets (Glickstein et al. 1972; Baker et al. 1976, Altman et al. 1976).

The author's own investigations show that in addition to the cerebellar afferents arising from the pontine nuclei proper and nucleus reticularis tegmenti, neurons in the corpus pontobulbare send axons to parts of the cerebellar cortex. The greatest number of neurons in this nucleus appear to project to the anterior lobe and paramedian lobule. A somewhat smaller number send axons to lobules VI–IX, crus I and II, and the paraflocculus. Although there do not appear to have been any previous studies on this pathway in the cat, or other infrahuman mammal, Marburg (1945) has described a projection from the corpus pontobulbare to the cerebellum in human material.

The name "corpus pontobulbare" has been used in the present study to describe the wedge-shaped aggregation of neurons which is found at the ventrolateral border of the caudal pontine tegmentum, just medial to the rootlets of the trigeminal nerve and the spinal trigeminal tract. This nucleus has commonly been believed to represent an "aberrant" pontine structure (Larsell and Jansen 1972). Accordingly, findings concerning projections from this nucleus to the cerebellum have been grouped together with results concerning the pontine nuclei proper and the nucleus reticularis tegmenti pontis. However, in a recent study Martin et al. (1977) argue that in the opossum the nucleus corporis pontobulbaris should be considered as a separate precerebellar nucleus on the basis of its afferent and efferent connections. Futhermore, this study suggests that neurons scattered further dorsally in the tegmentum around the rootlets of the trigeminal nerve at the level of the motor trigeminal nucleus (described in the present study simply as a "lateral tegmental cell group") actually comprise part of the nucleus corporis pontobulbaris. This matter is discussed in more detail below in the section concerning this so-called lateral tegmental cell group.

4.2 Raphe Nuclei

The raphe projection to the cerebellum was originally described by Brodal A et al. (1960) based on experiments using retrograde degeneration methods. Their findings suggested that the cerebellar projection arose from only three of the raphe nuclei (pontis pallidus, and obscurus) and that it terminated only in the intracerebellar nuclei (not in the cortex).

More recently, the advent of new histofluorescence and biochemical methods has provided evidence that the raphe nuclei supply the cerebellar cortex as well as the deep nuclei with serotonergic fibers. Histofluorescence studies in rat and cat show that the vast majority of the serotonergic neurons in the central nervous system are contained in the raphe nuclei (Dahlström and Fuxe 1964, rat; Pin et al. 1968, cat; Poitras and Parent 1978, cat). In the cat all of the raphe nuclei contain serotonergic neurons (Poitras and Parent 1978), although not all of the raphe neurons contain serotonin. Histofluorescence and biochemical studies have indicated the presence of sparsely distributed serotonergic terminals in the cerebellar cortex of the rat (Andén et al. 1967, Hökfelt and Fuxe 1969) in both molecular and granule cell layers. Likewise, studies in which tritiated serotonin has been injected intracisternally show that uptake of the labelled neurotransmitter occurs in axons within both the molecular and granule cell layers. According to Chan-Palay (1975,1976,1977) there are three distinct types of serotonergic afferents to the cerebellar cortex: (1) axons that terminate as classic mossy fiber rosettes in the granule cell layer, (2) axons that terminate diffusely throughout the cortical layers without specialized synaptic junctions, and (3) axons that traverse the molecular layer and bifurcate like parallel fibers. Unlike true parallel fibers, however, the latter do not terminate on Purkinje cell dendrites directly but rather on dendrites of cortical interneurons (Golgi cells, basket cells, and stellate cells).

Attempts to verify the existence of a raphe projection to the cerebellar cortex by anterograde autoradiographic tracing techniques after injections of tritiated amino acids into the raphe nuclei in rat and cat have yielded inconsistent results. Thus, Conrad et al (1974) reported negative findings in the cerebellum after tritiated proline injections into the dorsal and median (nucleus centralis superior) raphe nuclei in the rat. Taber Pierce et al. (1976) made no mention of fibers passing to the cerebellum after injections of tritiated proline into the dorsal raphe nucleus of the cat, however, negative findings are not specified. Since tritiated proline is not taken up and transported optimally by some cerebellar afferent systems (Künzle and Cuenod 1973) this is one possible factor in producing these negative findings. Another possibility is that the projection is simply too sparse to be visualized using tritiated amino acids. Bobillier et al (1976) found projections to the cerebellar cortex arising from the raphe nuclei centralis superior, pontis, and magnus, but not from the nucleus raphe dorsalis following stereotaxic injections of (^{14}C) leucine into these nuclei in the cat. The distribution of terminals within the cortex is not specified, however. Halaris et al. (1976) also obtained evidence of a projection to the cerebellar cortex in rats with injections of tritiated 5-hydroxytryptophan, leucine, or proline into the raphe nuclei dorsalis and centralis superior. In these experiments, however, transport of 5-hydroxytrytophan or amino acids was measured as total radioactivity per mg fixed brain tissue by liquid scintillation counting. Thus, again no data are available concerning the topographical distribution of raphe terminals within the cerebellar cortex. Clearly, further investigation will be necessary to provide information on this point.

Studies utilizing retrograde axonal transport of HRP support the notion that there is widespread raphe innervation of the cerebellar cortex. Shinnar et al. (1975) identified retrogradely labelled neurons in the raphe nuclei dorsalis (one labelled neuron only), centralis superior[1], magnus, pallidus, and obscurus after HRP injections into lobules VI and VII. Batini et al. (1978) also investigated afferents to lobules VI and VII and found clearcut evidence of a raphe projection only to lobule VII. They reported retrograde cell labelling in the raphe nuclei centralis superior, pontis, magnus, pallidus and obscurus.

It is worth noting that the distribution of cell-labelling in the raphe nuclei centralis superior and magnus in Fig. 6 of Batini et al. (1978) is much more extensive than that indicated by Shinnar et al. (1975) (their Fig. 2) or in a recent study by Taber Pierce et al. (1977). Also of interest is the fact that the labelled centralis superior neurons in Fig. 6 of Batini et al. (1978) appear to be distributed near the periphery of the nucleus. According to Poitras and Parent (1978) the serotonin-containing neurons are confined to the medial portion of the nucleus centralis superior, forming a single row on each side of the midline. Thus the neurons indicated by Batini et al (1978) would appear to mediate a nonserotonergic input to lobule VII.

The most comprehensive study of raphe nucleur projections to the cerebellar cortex is that of Taber Pierde et al. (1977). In this study the raphe nuclei were examined in a series of 55 cats with HRP injection sites in various parts of the cerebellar cortex. The results show that all parts of the cerebellar cortex except possibly lobule VI receive afferents from the raphe nuclei. The heaviest projection is to the vermis of lobules VIIA and X and to crus II. All of the raphe nuclei except the nucleus linearis intermedius and nucleus linearis rostralis project onto the cerebellar cortex. The raphe nuclei pontis and obscurus contain the largest number of neurons sending axons into the cerebellum.

---

1 Bernan's (1968) terminology. This is apparently equivalent to Taber's (Taber et al. 1960) raphe nuclei pontis and centralis superior. The labelled neurons in the study of Shinnar et al. (1975) appear to be mostly in the caudal part of this nucleus, corresponding to the nucleus raphe pontis of Taber et al.

Neurons in the nucleus raphe pontis project mainly to lobule VIIa, the nodulus, and crus II, and to a lesser extent to the lateral parts of lobules IV–V, the uvula, and the flocculus. Connections with lobules I – III, the medial portion of lobule V, lobules VIIB and VIII, crus I, and the paramedian lobule are apparently relatively sparse. There is no evidence of a projection from the nucleus raphe pontis to the medial part of lobule IV or the paraflocculus. Whether or not lobule VI receives a projection is unclear.

The largest number of neurons in the nucleus raphe obscurus project to the lateral part of lobule V and to crus II, with a slightly smaller proportion providing afferents to lobule VIIA and the uvula. A few neurons in the nucleus send fibers to lobules I–III, the medial and lateral parts of lobule IV, the medial part of lobule V, lobule VIIIB, crus I, the paramedian lobule, the paraflocculus, and the flocculus. Lobules VI, VIIB, VIIIA and the nodulus do not appear to receive a projection.

The cerebellar projection from the nucleus raphe magnus is mainly to the nodulus. A relatively small number of neurons in this nucleus give rise to axons terminating in the cortex of the anterior lobe, lobule VIIA, the uvula, crus II, the dorsal paraflocculus, and the flocculus.

Only a small number of neurons in the nucleus raphe pallidus give evidence of projecting to the cerebellum. These neurons send axons to the cortex of the anterior lobe, lobule VIIA, lobule VIIIB, the uvula, crus I and the paramedian lobule.

The dorsal raphe nucleus and nucleus centralis superior project mainly to the nodulus. A very small number of neurons in the nucleus centralis superior give evidence of projecting to the uvula and ventral paraflocculus. The nucleus raphe dorsalis also contains a scanty population of neurons sending axons to lobules I, II, the lateral part of lobule V, lobule VIIA, the uvula, the paramedian lobule, and the paraflocculus.

The results of the author's own experiments are slightly at variance with those of Taber Pierce et al. (1977). According to their study the heaviest raphe projection is to lobules VII and X of the vermis and to crus I. While the present study substantiates the finding of a relatively heavy raphe projection to the middle vermis, the evidence suggests that the projections to crus II and lobule X of the vermis are relatively minor. Of course, it is difficult to make quantitative comparisons between the two studies because it appears that in general the numbers of raphe neurons labelled after injections in various parts of the cerebellar cortex were much greater in the present study than in that of Pierce et al. This is likely to be due to procedural differences, possibly to differences in the amout of HRP injected or to variations in the placement of the injection site within a cerebellar region. In particular, the paucity of cell-labelling in the cases of Taber Pierce et al. (1977) with crus I injection sites may be due to the relatively small amounts of HRP solution injected compared to crus I injections in the present series. Another point of difference between the present study and that of Taber Pierce et al. (1977) concerns a raphe projection to the paraflocculus. In the present study there was no evidence of such a connection. However, according to their Table 5, the projection to the paraflocculus is quantitatively so small (only one or two cells per case) that it might not have been detected in the present study. A further point of variance concerns their finding a projection from the raphe nuclei dorsalis and centralis superior to the cerebellar cortex, in particular to lobule X. In the present study there was no evidence of a projection from the nucleus centralis superior to the cerebellum in any case. The nucleus raphe dorsalis contained labelled cells in only one case (CHRCB 20) with an injection site in the flocculus, and failure to confirm the finding in another case with a floccular injection site casts doubt on this finding. Possibly the absence of cell-labelling in these nuclei in case CHRCB 17 (uvula and nodulus injection site) can be attributed to a false negative, as is occasionally the case with the HRP method. The failure to observe cell-labelling in the raphe nuclei dorsalis and centralis superior in cases with superficial injection sites elsewhere in the cortex is not surprising, as Taber Pierce et al (1977) indicate that the projection of these nuclei to cerebellar cortical areas outside of lobule X is very minor.

Only scanty information is available concerning afferents to the nuclei which give rise to the raphe-cerebellar projection. The rostral-most cerebellar-projecting nuclei, the raphe nuclei dorsalis and centralis superior, may function as links in a circuit whereby limbic input is made available to the cerebellar cortex. A recent HRP study (Sakai et al. 1977) in the cat shows that the nucleus raphe dorsalis receives input from a variety of sources including the locus coeruleus complex, the parabrachial nuclei, the nucleus laterodorsalis tegmenti, the griseum centrale pontis, the substantia nigra, the lateral habenular nucleus, certain hypothalamic areas, preoptic areas, an area dorsolateral to the inferior olivary complex (and medial to the lateral reticular nucleus), and the other raphe nuclei (particularly the nuclei linearis intermedius, centralis superior, pontis, and magnus). In addition, the nucleus raphe dorsalis may receive a sparse input from certain large ganglion cells of the retina (Foote et al. 1978).

Anterograde degeneration studies have shown that the lateral preoptic region, the caudal part of the lateral hypothalamic region, and the habenular nuclei provide afferents to the nucleus centralis superior (Nauta 1958). In addition this nucleus receives a scanty projection from the fastigial nucleus, and from the cerebral cortex, particularly the sensorimotor region (Brodal A et al. 1960).

The sources of inputs to the caudal raphe nuclei are somewhat different from those which innervate the rostral nuclei. The afferents to the nucleus raphe pontis are apparently not known except for a projection from the fastigial nucleus to the dorsal part of the nucleus raphe pontis (Brodal A et al. 1960).

Anterograde degeneration studies show that fibers from the spinal cord terminate in the dorsal parts of the raphe nuclei magnus and pallidus, whereas fibers from the cerebral cortex (particularly the sensorimotor region) reach the ventral parts. The nucleus raphe magnus receives input from the fastigial nucleus (Brodal A et al. 1960) and from the head of the caudate nucleus (Usunoff et al. 1974). Autoradiographic evidence indicates that the nucleus raphe dorsalis provides afferents to the raphe nuclei magnus and obscurus (Taber Pierce et al. 1976). Apart from the input from the nucleus raphe dorsalis, the sources of afferents to the nucleus raphe obscurus in the cat appparently have not been described. However, a recent study provides some information on sources of afferents to the caudal raphe nuclei (magnus, obscurus, and pallidus) in the rat. According to the HRP study of Gallager and Pert (1978) the major sources of input to the caudal raphe nuclei are neurons in the mesencephalic central gray and the dorsal and ventral tegmental nuclei. A smaller proportion of the afferents arise in the inferior and superior colliculi (deep layers), medial vestibular nucleus, nucleus reticularis pontis caudalis, and nucleus reticularis gigantocellularis. Minor projections arise from the nucleus reticularis pontis oralis, and the cerebral cortex (particularly the ventromedial prefrontal cortex).

4.3 A Lateral Tegmental Cell Group at the Level of the Isthmus

The author's own investigations show that the cerebellum receives input from a group of neurons situated in the lateral tegmentum in and around the outgoing rootlets of the motor trigeminal and facial nerves at the level of the isthmus (see also Graybiel and Hartwieg 1974; Gould and Graybiel 1976; Faull 1977). This lateral tegmental cell group appears to project most strongly to the middle vermis and crus II. A smaller number of neurons in the lateral tegmentum project to anterior and posterior vermal regions, crus I, and the paramedian lobule.

Relatively little is known about the function of this lateral tegmental cell group, which appears to correspond in part to the intertrigeminal nucleus described by Taber (1961). Kuypers (1958) described a fiber projection from the face area of the cat's motor cortex which terminates in a zone of reticular formation interposed between the principal sensory nucleus of the trigeminus and the motor root of the trigeminus, which he termed a "juxta-trigeminal" region. Smith (1975) described an "intertrigeminal" commissural projection arising from a region of the parvicellular reticular formation lying between the principal sensory trigeminal nucleus and the motor trigeminal nucleus in the monkey.

Recent studies by Martin et al. (1977) and Kimoto et al. (1978) in the opossum and rat suggest that the neurons scattered around and among the emerging rootlets of the motor trigeminal nerve actually comprise part of the nucleus corporis pontobulbaris. This nucleus begins rostrally as a wedge-shaped aggregation of neurons at the ventral surface of the caudal pons, just medial to the spinal trigeminal tract and brachium pontis. (For a discussion of the cerebellar projections arising from this region in the present study see the section concerning the pontine nuclei). Further caually neurons of this nucleus are said to extend dorsally and medially among and around the fascicles of the motor trigeminal nerve. In the opossum, the nucleus corporis pontobulbaris appears to receive input from the facial motor-sensory cortex, the red nucleus, the spinal cord, and the cerebellum. Experiments employing the HRP technique reveal that this nucleus also projects to widespread areas of the cerebellar cortex in the opossum and rat.

It is not yet entirely clear whether the precerebellar neurons lying around and among the rootlets of the motor trigeminal nerve should be considered as part of the nucleus corporis pontobulbaris (as suggested by Martin et al. 1977), whether they should be considered as part of the adjacent trigeminal complex, or whether they constitute a more or less independent cell group. However, recent evidence favors the notion that these lateral tegmental neurons comprise a more or less independent cell group concerned in some way with supranuclear control of oculomotor function. According to Graybiel and Hartwieg (1974) neurons in this lateral tegmental cell group (but apparently not in the wedge-shaped nucleus designated as the nucleus corporis pontobulbaris in the present study) project to the region of the oculomotor complex in the cat. Within the lateral tegmental area neurons giving rise to the oculomotor pathway are arrayed in a distribution which is strikingly similar to the distribution of neurons projecting to the cerebellar cortex (cf. the distribution of neurons indicated by arrowheads in Graybiel and Hartwieg (1974), Figs. 1 and 2; and the present study, Figs. 1 – 3, 6 (CHRCB 13 only), and 8). Neurons in this tegmental area also receive input from the region of the oculomotor and trochlear nuclei (see Fig. 14 of Graybiel 1977). Mehler (1969) and Faull (1978) have reported that, in the rat, the cerebellum projects back to the lateral tegmental cell group by way of the uncrossed descending brachium conjunctivum, confirming an earlier description of this projection by Ramon y Cajal (1903). Thus, neurons in this lateral tegmental cell group maintain reciprocal connections with both the oculomotor nuclei and the cerebellum. Although it cannot yet be determined whether the cerebellar and oculomotor pathways are mediated by the same neurons, the evidence does suggest that this part of the lateral tegmentum gives rise to an oculomotor side-path through the cerebellum by which the cerebellum could exercise supranuclear control over eye movements.

4.4 Locus Coeruleus

Histofluorescence techniques provided the first evidence of a catecholaminergic input to the cerebellar cortex. Following lesions of the cerebellum in rats Anden et al. (1967) observed swollen fluorescent cell bodies indicative of retrograde changes in the A4 group of Dahlström and Fuxe (1964) and the locus coeruleus. Later studies indicated that the noradrenergic fibers enter the cerebellum by way of the inferior cerebellar peduncle, terminating in a sparse, patchy pattern over pratically all of the cerebellar cortex and in the deep nuclei. Within the cerebellar cortex the terminals seemed to make axodendritic contacts in both the molecular and granular layers (Hökfelt and Fuxe 1969). Furthermore, lesions of the dorsal neuronal pathway to the cerebral cortex resulted in increased levels of amine in the cerebellar cortex, presumably due to increased amine transport from locus coeruleus noradrenergic cell bodies to the remaining intact fibers of cells whose ascending axons had been severed. This finding suggests that cerebellar noradrenergic fibers arise from neurons which also send axons to the cerebral cortex (Olson and Fuxe 1971).

Histofluorescence studies have also provided evidence for a noradrenergic innervation of the cerebellar cortex of the cat. According to Maeda et al. (1973) the noradrenergic cerebellar fibers arise as a branch of the dorsal descending pathway from the locus coeruleus and enter the cerebellum with the brachium pontis. Some of the cerebellar-projecting neurons also appear to send branches to the cerebral cortex (Chu and Bloom 1974). In the cerebellar cortex, the noradrenergic fibers terminate sparsely in the molecular layer and around the Purkinje cell somata (Chu and Bloom 1974).

It is not yet clear whether the locus coeruleus is the sole source of the cerebellar noradrenergic innervation in rat and cat or, conversely, whether the entire projection from the locus coeruleus to the cerebellum is noradrenergic. Histofluorescence studies indicate that in the rat all, or virtually all, of the locus coeruleus neurons contain norepinephrine (Dahlström and Fuxe 1964) so that the projections from the locus coeruleus proper are probably entirely noradrenergic. A recent study using the HRP method combined with monoamine oxidase staining supports this notion (Kimoto et al. 1978). However, although the cerebellar catecholamine innervation is mainly supplied by the locus coeruleus (Tohyama 1976), noradrenergic neurons in the nucleus subcoeruleus and possibly the nucleus dorsalis nervi vagi and the nucleus commissuralis also project to the cerebellar cortex in the rat. Neurons in other areas of the rat brain stem which contain noradrenergic neurons project to the cerebellum as well, as evidenced by their retrograde labelling with HRP. These cerebellar-projecting cells may not be noradrenergic, however, as they do not show evidence of monoamine oxidase activity. (This includes some of the neurons in the nucleus subcoeruleus).

Slightly less information is available in the cat. In contrast to the situation in rat, where the locus coeruleus exists as a well-defined group of densely packed norepinephrine-containing neurons, the catecholaminergic neurons in the cat locus coeruleus are mingled with nonnoradrenergic neurons. Moreover, the catecholaminergic neurons are rather diffusely distributed in the nucleus subcoeruleus, the parabrachial nuclei, and the so-called Kölliker-Fuse nucleus in the cat, which also contain nonnoradrenergic neurons (Pin et al. 1968; Chu and Bloom 1974; Jones and Moore 1974; Poitras and Parent 1978). According to Chu and Bloom (1974) neurons in the locus coeruleus proper are not the major source of cerebellar norepinephrine in the cat. Rather, noradrenergic cells in the parabrachial nuclei, the Kölliker-Fuse nucleus, and in the location corresponding to Dahlström and Fuxe's (1964) group A4 contribute the major cerebellar norepinephrine innervation.

There are relatively few studies of the locus coeruleus-cerebellar projection using anterograde tracer techniques. Following injections of tritiated proline into the locus coeruleus in rats Pickel et al. (1974) found that locus coeruleus fibers enter the cerebellum by way of the superior cerebellar peduncle and terminate most heavily around Purkinje cell somata and in the molecular layer in the anterior lobe and flocculus, and to a lesser extent in the cerebellar neocortex. Histofluorescence studies which reveal proliferation of norepinephrine containing axons in the rat cerebellar cortex after partial lesions of the superior cerebellar peduncle also provide support for the notion that locus coeruleus axons enter the cerebellum by way of the superior cerebellar peduncle in the rat (Pickel et al. 1973). A recent autoradiographic study in the monkey indicates only that locus coeruleus fibers enter the cerebellum ventral to the deep nuclei (Bowden et al. 1978).

It is as yet unclear how widely the locus coeruleus fibers are distributed within the cerebellar cortex. Two recent studies using the HRP method provide some information on this point. According to Kimoto et al. (1978) noradrenergic locus coeruleus neurons innervate the entire cerebellar cortex of the rat bilaterally, while noradrenergic nucleus subcoeruleus neurons project to the cerebellar neocortex. (Neurons in the nucleus subcoeruleus which do not show monoamine oxidase activity, and therefore are probably not noradrenergic, also project to the anterior lobe). An HRP study in the cat (Somana and Walberg 1978b) indicates that fibers arising from cell bodies in the caudal half of the locus coeruleus proper project to the entire cerebellar vermis (most strongly to its anteriormost and posteriormost parts) and to the flocculus and ventral paraflocculus. (Findings concerning possible projections from the nucleus subcoeruleus and parabrachial nuclei are not mentioned). There was no evidence of a projection to other parts of the hemispheral cerebellar cortex. Possibly this is a falsely negative finding due to the limitations of the technique, as Kimoto et al. (1978) have indicated that there is a substantial locus coeruleus projection to the hemispheral cerebellar neocortex in the rat. Alternatively, this negative finding could reflect a real species difference.

Horseradish peroxidase studies in the rat and cat indicate that afferents to the locus coeruleus arise from a variety of sources including parts of the cerebral cortex; the amygdala; the diencephalic central gray; the parafascicular nucleus; the principal sensory and spinal trigeminal nuclei; the vestibular nuclei; the lateral reticular nucleus; the parasolitary nucleus; the nucleus commissuralis; the nucleus praepositus hypoglossi (projections from these nuclei have been described in rat only); the hypothalamic nuclei (including the preoptic area); the raphe nuclei; the mesencephalic central gray; areas of the mesencephalic, pontine, and medullary reticular formation; the parabrachial nuclei; the nucleus of the solitary tract; the fastigial nucleus (rat and cat); the nucleus laterodorsalis tegmenti and the substantia nigra (the latter two are described in cat only) (Sakai et al. 1977; Cedarbaum and Aghajanian 1978). Anterograde degeneration studies also provide evidence of a direct projection to the anterior part of the locus coeruleus in the cat from the fastigial nucleus (and possibly Purkinje cells in the vermis) by way of the brachium conjuctivum (Snider 1975).

The functional significance of the locus coeruleus-cerebellar projection remains to be elucidated. It is worthy of note that the noradrenergic afferents to the cerebellar cortex appear to synapse mainly on Purkinje cells, in contrast to the serotonergic afferents, which apparently synapse entirely on interneurons (Bloom et al. 1971).

The author's own in investigations provide no information on the locus coeruleus projection to the cerebellar cortex due to the fact that in the present material endogenous peroxidase activity was often observed in the catecholaminergic cell groups of the dorsolateral pontine tegmentum which was difficult to distinguish from HRP reaction product attributable to retrograde transport of exogenous tracer substance.

4.5 Trigeminal Nuclei

Prior studies have indicated the existence of secondary trigeminal projections to the cerebellum in several mammalian species. Woodburne (1936) described fibers from the principal sensory nucleus of the trigeminus and the spinal trigeminal complex passing to the cerebellum in the rabbit. Larsell (1947) saw fibers from the superior sensory nucleus going into the cerebellum in human embryos. Although Torvik and Brodal (unpublished study cited in Brodal A 1954) were unable to confirm the existence of a trigemino-cerebellar pathway in retrograde degeneration experiments in kittens, later experiments by Carpenter and Hanna (1961), using anterograde degeneration methods, established that the spinal trigeminal nucleus, pars oralis and pars interpolaris, projects to lobules V and VI of the cerebellar cortex in the cat. However, the pars caudalis of the spinal trigeminal nucleus did not appear to send fibers to the cerebellum (Stewart and King 1963). Karamanlidis (1968) found retrograde changes in the principal sensory nucleus of the trigeminus and in the pars oralis and pars interpolaris of the spinal trigeminal nucleus of the goat after large lesions of the cerebellum.

More recently, experiments employing anterograde and retrograde tracer techniques have provided additional evidence of trigeminocerebellar projections in rat, cat, and mouse. According to Steindler (1977) afferents arising in the principal sensory trigeminal nucleus and spinal trigeminal nucleus, pars oralis and interpolaris, pass to vermal and hemispheral cerebellar cortex in the mouse. Similar findings have been reported in the rat (Watson and Switzer 1978) except that in this species the pars oralis of the spinal trigeminal complex appears to contain only a few neurons projecting to cerebellar cortex, and these are localized in the dorsomedial tip of the nucleus. Autoradiographic experiments (Courville and Faraco-Cantin 1978) provide some evidence that fibers from the spinal trigeminal nucleus terminate as mossy fibers in the anterior lobe and lobulus simplex in the cat.

The most complete study of trigeminocerebellar connections in the cat is that of Ikeda (1979), based on experiments using the HRP technique. According to Ikeda, neurons in the spinal and principal trigeminal nuclei project to the dorsal part of the paramedian lobule, the posterior folia of crus II, lobulus simplex, and lobules V-VIIIa. The trigeminocerebellar neurons are found mainly in the pars interpolaris and caudal third of the pars oralis of the spinal trigeminal nucleus, and more rostrally in the rostralmost part of the pars oralis and in the ventral part of the principal sensory nucleus. A few neurons in the subnucleus magnocellularis of the pars caudalis of the spinal trigeminal nucleus also project to these cerebellar cortical areas. There appears to be a rather loose topographical organization of the projections from the pars oralis and pars interpolaris of the spinal trigeminal complex to the hemispheral cortex, so that neurons projecting to the paramedian lobule and lobulus simplex are located more dorsally, while those projecting to crus II are located more ventrally within the spinal trigeminal complex. There was no evidence of trigeminal projections to the anterior portion of the anterior lobe, crus I, the anterior folia of crus II, the paraflocculus, flocculus, or the ventral part of the paramedian lobule. The findings in the author's own investigations are in all respects compatible with those of Ikeda concerning cerebellar projections arising from the spinal trigeminal and principal trigeminal nuclei.

Other investigations have provided evidence of the existence of a mesencephalic trigeminocerebellar pathway in the rat (Cupedo 1965; Eller and Chan-Palay 1976), cat (Brodal A and Saugstad 1964), monkey (Chan-Palay 1977), and human (Pearson 1949).

It is unclear whether mesencephalic trigeminal fibers terminate in the cerebellar *cortex* in most species, however, as only one of these studies (Chan-Palay 1977) has provided evidence that mesencephalic trigeminal axons reach the cerebellar cortex. The report by Eller and Chan-Palay (1976) suggests that these fibers terminate, at least in part, in the lateral cerebellar nucleus in the rat. In the monkey, the mesencephalic trigeminal nucleus apparently projects both to the deep cerebellar nuclei (dentate and posterior interposed nuclei) and to the overlying cortex (lobule IVa; Chan-Palay 1977).

Other HRP studies have provided evidence that the motor trigeminal nucleus sends fibers to the cerebellum. The motor trigeminal nucleus projects to the lateral cerebellar nucleus in rat and monkey (Eller and Chan-Palay 1976; Chan Palay 1977). However, a recent report by Kotchabhakdi and Walberg (1977) shows that motor trigeminal fibers also reach lobules I and II of the cerebellar cortex in the cat and monkey.

In the present study there was no evidence of cerebellar cortical projections arising from the mesencephalic or motor trigeminal nuclei (note that lobules I and II were not investigated, however) except in a single case with an injection site centered in the flocculus. The trigeminal findings in this case must be considered doubtful due to the possible involvement of the trigeminal nerve in the injection site.

4.6 Vestibular Nuclei

Prior studies employing anterograde and retrograde degeneration techniques suggested that both primary and secondary vestibular afferents reach only limited ventral regions of the cerebellum. Brodal A and Hoivik (1964) found that primary vestibular fibers terminate ipsilaterally in the flocculonodular lobe, uvula, lingula, dorsal and ventral paraflocculus, and in the small-celled part of the dentate nucleus. Similarly, Brodal A and Torvik (1957) reported that secondary vestibulocerebellar projections are distributed only to the flocculus, nodulus, uvula, and possibly to the paraflocculus in the cat. These secondary vestibular afferents were found to arise from the caudal parts of the medial and descending vestibular nuclei (including cell group f) and cell group x.

More recently, experiments using the HRP method have shown that the distribution of secondary vestibular afferents within the cerebellar cortex is somewhat more widespread than previously believed. Several studies in which HRP was injected into various parts of the cerebellar vermis outside of the traditional vestibulocerebellum showed that these cortical areas also receive input from the brain stem vestibular complex (Fig. 2 of Shinnar et al 1975; Gould and Graybiel 1976; Precht et al. 1977; Batini et al. 1977). Furthermore, the cerebellum receives afferents from several vestibular cell groups which had not previously been thought to give rise to vestibulocerebellar fibers: the superior vestibular nucleus, cell group y, the supravestibular nucleus, and possibly the lateral vestibular nucleus (Gould and Graybiel 1976; Mehler 1977; Kotchabhakdi and Walberg 1978b).

The most comprehensive study of the vestibulocerebellar projection in the cat by HRP methods is that of Kotchabhakdi and Walberg (1978b). They found that the entire cerebellar vermis and the flocculus receive input from various parts of the vestibular complex. The medial and descending vestibular nuclei consistently contained retrogradely labelled neurons after injections into various parts of the vermis and the flocculus. The superior and lateral vestibular nuclei also contained HRP-positive neurons, bilaterally, after injections in the flocculus and nodulus or after large injections in the vermis. In addition, these studies provided evidence that cell groups f and x and the supravestibular nucleus project to parts of the vermis, while cell group y projects mostly to the caudal half of the vermis and the flocculonodular lobe. There was no evidence that secondary vestibular fibers pass to crus I and II, the paramedian lobule, or the paraflocculus.

Some discrepancies exist between the author's own findings and those of Kotchabhakdi and Walberg (1978b). In the present study there was no evidence of a projection from the lateral vestibular nucleus to the cerebellar cortex. Since this is apparently a very minor projection (see Kotchabhakdi and Walberg 1978b) possibly the methods used in the present study would not have revealed it. However, Rubertone and Haines (1979) also reported negative findings in the lateral vestibular nucleus after HRP injections in the flocculus, nodulus, uvula, or paraflocculus of the lesser bushbaby. Similarly, Yamamoto (1979) reported cell-labelling in the superior, medial, and descending vestibular nuclei after HRP injections in the flocculus of the rabbit, but made no mention of labelling in the lateral vestibular nucleus. As to the superior vestibular afferents to the cerebellum, the results of the present experiments show that there is a fairly strong projection to the flocculus and nodulus, but there is no evidence of a projection to the rest of the vermis. Batini et al. (1978) also failed to find labelled neurons in the superior and lateral vestibular nuclei following HRP injections into lobules VI and VII. Again, according to Kotchabhakdi and Walberg (1978b) the projection from the superior vestibular nucleus to the vermis outside of the nodulus is quantitatively a minor one and only demonstrable in cases with fairly large injection sites (they do not specify how much HRP solution was injected) so that possibly it would not be revealed by the procedures used in the present study or that of Batini et al. (1978).

In contrast to the findings of Kotchabhakdi and Walberg (1978b) the author's investigations in the cat and those of Kimoto et al. (1978) in the rat provide evidence that the vestibular nuclei may also project to parts of the hemispheral cortex. In the author's experiments, HRP injections centered in crus I or II elicited retrograde labelling in small numbers of neurons in cell group f and the medial and descending vestibular nuclei. A fair number of neurons in these nuclei are also labelled in CHRCB 10, with an injection site in the paramedian lobule. In addition, there is evidence of a modest projection from cell group x to the paramedian lobule. The vestibular cell-labelling in cases with injection sites in crus I or the paramedian lobule might be accounted for by slight spread of the injected HRP solution to areas which are now known to receive a vestibular projection (e.g., lobulus simplex and lobule VII). This explanation would not account for the sparse vestibular cell-labelling in the case with a crus II injection site, however, as in this case the HRP solution has spread only to adjoining parts of crus I and the paraflocculus. Thus, the author's findings suggest the possibility that the vestibular nuclei send a sparse projection to the medial parts of the hemisphere, as is apparently true of the projections from the lateral reticular nucleus (see Künzle 1975). Were this the case, secondary vestibular afferents to the hemispheral cerebellar cortex might be demonstrable only in cases with large HRP injection sites encroaching on the medial border of the hemisphere.

The distribution of the primary vestibulocerebellar projection has also been reinvestigated by Kotchabhakdi and Walberg (1978a) by the HRP technique. Primary vestibulocerebellar fibers are found by this method to be distributed to the entire vermis. A recent study by Korte and Mugnaini (1979) using anterograde degeneration methods provides somewhat different results. The anterograde studies suggest that primary vestibular afferents terminate mostly in the uvula and nodulus, with a sparse projection to the flocculus (apparently not investigated by Kotchabhakdi and Walberg 1978a). There was no evidence of a primary vestibular projection to the cerebellar vermis outside of the uvula and nodulus. However, the anterograde method is not ideally suited for detecting what may be a very sparse projection.

These new findings concerning the distribution of primary and secondary vestibular afferents to the cerebellar cortex suggest that the old notion of a "vestibulo-cerebellum" consisting solely of the ventral parts of the cerebellar cortex is no longer tenable. Moreover, reexamination of the afferent organization of the vestibular nuclei shows that the superior, medial, and descending vestibular nuclei and cell group y receive primary vestibular input over their entire areas (Korte 1979). Of the cerebellar-projecting vestibular nuclei, only cell groups f and x do not appear to receive primary vestibular afferents. Thus, in contrast to prior studies which suggested that the cerebellar-projecting nuclei relayed mainly spinal input to the cerebellar cortex (Walberg et al. 1958; Gacek 1969) the findings in the most recent studies suggest that the vestibular nuclei may convey mainly vestibular information.

The cerebellar-projecting parts of the vestibular nuclei receive afferents not only from the vestibular ganglion, but also from the spinal cord, the cerebellar flocculonodular lobe and fastigial nucleus, the red nucleus and the vestibular nuclei of the opposite side. Commissural fibers arise from neurons in cell group y, and the superior and medial vestibular nuclei, and to a lesser extent from the descending nucleus (Pompeiano et al. 1978; Gacek 1978) and terminate mostly in the corresponding nuclei on the opposite side (Ladpli and Brodal A 1968). Fibers from the spinal cord end mainly in cell group x and at the caudal poles of the descending and medial vestibular nuclei (Pompeiano and Brodal A 1957; Brodal A and Angaut 1967). Cell groups f and x receive afferents from the red nucleus (Edwards 1972) as well as from the cerebellar flocculonodular lobe and the fastigial nucleus (Angaut and Brodal A 1967; Walberg et al. 1962). (It is of some interest that cell group f, which receives input from the flocculondular lobe and the fastigial nucleus, does not appear to project very strongly to these areas (Ruggiero et al. 1977). Rather, the major projection from cell group f is to the vermis rostral to the nodulus and uvula.) The flocculus and nodulus also project on cell group y (Haines 1977, Galago; Gould 1979, cat).

4.7 Perihypoglossal Nuclei

Experimental studies by Brodal A (1952) and Torvik and Brodal A (1954) first established that the perihypoglossal nuclei, consisting of the nucleus intercalatus of Staderini, the nucleus praepositus of Marburg, and the nucleus of Roller, project onto the cerebellum of the cat. All three of the perihypoglossal nuclei appeared to provide fibers to the cerebellar cortex which were distributed uniformly throughout the anterior lobe, and to lobules VIII and IX of the posterior vermis, but not to the cerebellar hemisphere.

Recent HRP studies confirm and extend these prior findings. The most comprehensive study of the perihypoglossal projection to the cerebellum is that of Kotchabhakdi et al. (1978). These authors confirm earlier reports that all three of the perihypoglossal nuclei send fibers to the cerebellum. However, the distribution of perihypoglossal afferents within the cerebellar cortex is more widespread than previously described. Fibers from the nuclei intercalatus and praepositus hypoglossi reach the entire vermis, the flocculus, and the paraflocculus. The nucleus intercalatus appears to project more strongly to the anterior vermis, while the nucleus praepositus hypoglossi projects more to the posterior vermis. Furthermore, the posterior vermis receives afferents from neurons located mostly in the rostral part of the nucleus praepositus hypoglossi and nucleus intercalatus, while the anterior vermis receives afferents from more neurons in the caudal halves of these nuclei. Fibers from the nucleus of Roller appear to be more restricted in their distribution, supplying only the anterior lobe vermis, lobules VI and VII, and the ipsilateral flocculus. According to Kotchabhakdi et al. (1978) there is no evidence of a projection from the perihypoglossal nuclei to crus I and II or the paramedian lobule.

Findings from other studies are largely in agreement with those of Kotchabhakdi et al. (1978). Batini et al. (1977) observed cell-labelling bilaterally in all three peri-hypoglossal nuclei after HRP injections into lobules VI and VII. Shinnar et al. (1975) and Gould and Graybiel (1976) noted the presence of HRP-positive neurons in the nucleus praepositus hypoglossi after injections in lobules VI – VII and lobules V – VII respectively, but did not mention findings in the other nuclei. Alley et al. (1975) noted bilateral cell-labelling in all three nuclei following HRP injections into lobule VI, the nodulus, and the flocculus in rabbits. The evidence concerning the projection from the nucleus of Roller is thus slightly at variance with the findings of Kotchabhakdi et al. (1978). Whereas Kotchabhakdi et al. (1978) found no indication of a projection from the nucleus of Roller to lobules VIII – X, and only an ipsilateral projection to the flocculus, Alley et al. (1975) reported a bilateral projection from the nucleus of Roller to both the flocculus and the nodulus in the rabbit. Findings in the author's own studies in the cat in cases with HRP injection sites concerned in the flocculus and in lobules VIII – X are in agreement with Alley et al. (1975) in providing evidence of a bilateral projection from the nucleus of Roller to the flocculus and caudal vermis.

The author's investigations also suggest the existence of a sparse projection from the perihypoglossal nuclei to crus I and II and the paramedian lobule. In the present study a modest number of retrogradely labelled neurons were found in all subdivisions of the perihypoglossal nuclei in cases with superficial injection sites centered in crus II and the paramedian lobule. Somewhat fewer HRP-positive neurons were encountered in the cases with injection sites centered in crus I. These findings might be explainable by the extension of the crus I injection sites into lobules simplex or the inclusion of a small part of the dorsal paraflocculus with the crus II injection site, since Kotchabhakdi et al. (1978) have shown that both the paraflocculus and lobulus simplex receive a sparse projection from the perihypoglossal nuclei. The findings after an HRP injection into the paramedian lobule are more difficult to explain, as cell-labelling in the perihypoglossal nuclei is fairly extensive, and there is only very minor extension of the injection site into lobule VII in this case. According to Kimoto et al (1978), HRP injections in the hemispheral cerebellar cortex in the rat also elicit cell-labelling in substantial numbers of neurons in the perihypoglossal nuclei. Their Tables 2 and 4 indicate that the retrograde cell-labelling in the nucleus praepositus hypoglossi is not dependent on inclusion of lobulus simplex in the injection site. While it is possible in both these studies that the perihypoglossal cell-labelling in cases with hemispheral injection sites is produced by spread of the injection site into areas which have been shown to receive a projection from these nuclei, the possibility that these nuclei send a sparse projection to the hemisphere must also be considered. Satisfactory resolution of this issue will require further investigation of the connections of the perihypoglossal nuclei by autoradiographic methods.

In light of the known anatomic connections of the perihypoglossal nuclei, it seems likely that these cell groups mediate another oculomotor side-path through the cerebellum (Baker and Berthoz 1975) comparable to those suggested for the lateral tegmental cell group (see discussion above) and the vestibular nuclei (see Ito 1972). Recent HRP studies in the cat have demonstrated that the nucleus praepositus hypoglossi sends fibers to the oculomotor (Graybiel and Hartwieg 1974; Gacek 1977), trochlear (Gacek 1979b), and abducens nuclei (Maciewicz et al. 1977; Gacek 1979a). As to the afferent connections of the perihypoglossal nuclei, Baker and Berthoz (1975) have shown by electrophysiologic methods that the perihypoglossal nuclei receive input from secondary vestibular neurons. The perihypoglossal nuclei also receive a projection from the fastigial nucleus (Walberg 1961) and the flocculonodular lobe (Angaut and Brodal 1967). Afferents to the perihypoglossal nuclei arise, additionally, from the motor cortex, particulary the face area (Kuypers 1958; Sousa-Pinto 1970).

It is important to note that the perihypoglossal oculomotor side-path passes not only to the classic vestibulocerebellum, but also to widespread areas of the vermis and possibly to other parts of the hemisphere. Stimulation studies in cats and monkeys (Cohen et al. 1965; Ron and Robinson 1973) have shown that there are at least three cerebellar cortical areas concerned with eye movements: the flocculonodular lobe, crus I and II, and a midvermal area. The perihypoglossal nuclei provide input to at least two of these regions.

4.8 Inferior Olive

It is now widely believed that at least the majority (if not all) of the cerebellar climbing fibers arise from neurons in the inferior olive (see, for example, Desclin 1974; Batini 1976; Courville and Faraco-Cantin 1978). However, even before the mode of termination of the olivocerebellar projection was described, the topographical organization had been mapped in considerable detail in the cat by Brodal (1940) using retrograde degeneration methods. These early studies established that all parts of the inferior olive project to the cerebellum and, conversely, that all areas of the cerebellar cortex receive olivary afferents. The connections from the inferior olive appear to be entirely crossed.

Using retrograde degeneration methods the pattern of olivocerebellar localization appeared remarkably sharp, with each cortical region receiving input from one or two subdivisions of the olive. According to this scheme the vermis proper of the anterior lobe receives fibers from the lateral dorsal accessory olive throughout its rostrocaudal extent and from the lateral part of the caudal medial accessory olive. The lateral parts of the anterior lobe are supplied by the medial parts of the dorsal accessory olive. Lobules VI – VIII are projected on by the medial part of the caudal medial accessory olive. The uvula receives input from the dorsomedial cell column and nucleus β. Adjoining medial and lateral segments of the medial accessory olive at middle levels project on the nodulus and flocculus respectively. The paramedian lobule is supplied by caudal parts of the ventral lamella of the principal olive. Rostral regions of the ventral lamella supply crus II, while the dorsal lamella of the principal olive supplies crus I. The paraflocculus receives input from the region of the ventrally located junction between the dorsal and ventral lamellae of the principal olive.

Recent investigations using the HRP method (Brodal A et al. 1975; Brodal A 1976; Hoddevik et al. 1976; Brodal A and Walberg 1977a,b; Hoddevik and Brodal A 1977; Kotchabhakdi et al. 1978) and the autoradiographic method (Groenewegen and Voogd 1977; Groenewegen et al. 1979; Courville and Faraco-Cantin, in press) demonstrate that the olivocerebellar projection is somewhat more complex than previously described, but support certain basic principles of organization elucidated by the earlier studies. Most notably, these recent studies show (in agreement with Brodal A 1940) that projections to the vermis proper arise mostly from neurons located in the caudal half of the inferior olive, while fibers passing to the hemispheral cortex derive from the rostral inferior olive (cf. also Figs. 2 and 3 of this paper).

The HRP studies show that the anterior lobe is organized into several longitudinal zones whose olivary afferents differ (Brodal A and Walberg 1977a). The middle region of the vermis proper receives input from the caudal medial accessory olive. The lateral part of the vermis proper receives input from the lateral half of the dorsal accessory olive. An intermediate zone in lobules IV-V is supplied by the medial part of the dorsal accessory olive and by a cell group in the rostral medial accessory olive. The lateral parts of lobules IV-V receive input from the dorsal lamella of the principal olive. A similar scheme has been proposed by Groenewegen and Voogd (1977) and Groene-wegen et al. (1979) based on analysis of autoradiographic material.

Experiments using the HRP method have not provided evidence of a longitudinal subdivision of lobules VI-VIII (Hoddevik et al. 1976). These studies indicate that each lobule receives afferents from largely separate regions in the caudal half of the medial accessory olive. Lobule VIII receives input from neurons in the lateralmost part of the medial accessory olive, while neurons in the medialmost part project to lobule VII. The middle (from medial to lateral) part of the caudal medial accessory olive sends fibers to lobule VI. More recent autoradiographic studies suggest, however, that an intermediate strip of cortex in lobules VI and VIII may be projected on by the caudal dorsal accessory olive, and that the lateralmost portions of these lobules receive projections from the rostral part of the inferior olivary complex (Courville and Faraco-Cantin in press).

Hoddevik et al. (1976) also indicate that there is a weak projection from the nucleus β to lobule VIIB, but such a projection was not evident in the author's material in cases with superficial injection sites (no spread of the injected HRP solution to the central white matter or the deep nuclei) in lobules V-VII. Since the nucleus β apparently supplies afferents to the fastigial nucleus (Hoddevik et al. 1976; Ruggiero et al. 1977) it seems likely that the cell-labelling reported by Hoddevik et al. (1976) in the nucleus β after injections in lobule VIIB is due to uptake by fibers passing to the fastigial nucleus. Although cell-labelling was present in the nucleus β even in cases in which the injection site did not encroach on the fastigial nucleus, it is possible that the injected HRP solution might have been taken up by axons passing through the central white matter.

Both HRP and autoradiographic experiments provide evidence of a longitudinal organization of the olivary projection to the uvula, although there is some disagreement on the details. The HRP studies (Brodal A 1976) indicate that the dorsomedial cell column and nucleus β project onto the medial zone of the uvula, while clusters of neurons in the rostral and caudal medial accessory olive project to the lateral zones of this lobule. One recent autoradiographic study suggests a different arrangement, however. According to Courville and Faraco-Cantin (in press) the caudal medial accessory olive and nucleus β project on a medially located strip in the uvula, while the lateral part of the uvula is projected on by cells in the rostral part of the inferior olivary complex (including the dorsomedial cell column).

A recent report by Phelan and Mehler (1979), based on HRP studies in the cat, shows that the nodulus receives afferents from the dorsal cap, the ventrolateral outgrowth, and the rostral part of the medial accessory olive. These findings are similar to those reported by Hoddevik and Brodal A (1977) after HRP injections into the nodulus in rabbits. It should be noted, however, that in all cases reported by both sets of authors, there has been some spread of HRP at the injection site into the adjoining parts of the uvula, with retrograde cell-labelling occurring to some degree also in the nucleus β and dorsomedial cell column. Although these two regions of the olivary complex are known to project to the uvula (Brodal A 1976), one cannot, on the basis of the available evidence, eliminate the possibility that the nucleus β and dorsomedial cell column also project to parts of the nodulus. The results in the author's case CHRCB 17, with an injection site involving the ventral part of the uvula and the adjoining part of the nodulus, support this notion. In this case HRP-positive neurons are found in the dorsomedial cell column, nucleus β, and parts of the medial accessory olive, but not in the dorsal cap or ventrolateral outgrowth.

Similar findings have been reported by Linauts and Martin (1978) after injections into comparable cerebellar cortical regions in the opossum.

The projection to the paramedian lobule has recently been shown to arise from four circumscribed areas of the inferior olive, including the caudal part of the ventral lamella of the principal olive (confirming Brodal 1940), the dorsal accessory olive, the rostral part of the medial accessory olive, and the caudal part of the dorsal lamella of the principal olive (Brodal A 1975). The olivocerebellar projection is organized in such a way that each of four longitudinal zones in the paramedian lobule receives input from a separate subdivision of the inferior olive (Brodal A and Walberg 1977b; Walberg and Brodal A 1979). Thus, the medialmost longitudinal zone receives afferents from the dorsal accessory olive; the laterally adjacent strip of cortex is projected on by the rostral medial accessory olive; fibers from the dorsal lamella of the principal olive pass to the next most lateral zone of the paramedian lobule; while the lateralmost strip of the middle folia receives afferents from the ventral lamella.

Afferents to lobulus simplex, crus I and crus II arise from the principal olive, the rostromedial part of the dorsal accessory olive, and the rostral part of the medial accessory olive (Kotchabhakdi et al. 1978). In agreement with Groenewegen et al. (1979) the HRP studies provide evidence that each part of the inferior olivary complex projects preferentially on a particular longitudinal zone involving parts of crus I, crus II and lobulus simplex. Afferents to the lateral parts of these lobules derive from the principal olive. Medial parts of the dorsal and ventral lamellae project to crus I and lobulus simplex, while the lateral bend region of the inferior olive projects to crus II. Zones at the medial and lateral edges of the medial half of these lobules receive afferents from the dorsomedial part of the dorsal accessory olive. The intermediate area in the medial half of crus I, crus II, and lobulus simplex receives fibers from the rostral part of the medial accessory olive.

The author's findings in cases with injection sites in lobulus simplex, crus I and crus II are largely in agreement with those of Kotchabhakdi et al. (1978) except that in the present material there is no clear distinction between regions of the principal olive projecting to crus I and crus II. Also, although Kotchabhakdi et al. (1978) observed cell-labelling in the ventrolateral outgrowth in two cases with injection sites in crus I and crus II, respectively, they attributed this finding to spread of HRP solution to lateral parts of lobule V or the central white matter (where it could presumably be taken up by fibers passing to the flocculus). This explanation does not seem to account for the cell-labelling in the dorsal cap in cases with crus I injection sites in the present series. HRP-positive neurons were found in the dorsal cap in all four cases with injection sites in crus I and in the ventrolateral outgrowth in two of the cases. Since there was no spread of HRP solution to the central white matter in these cases, uptake by fibers of passage does not seem likely to explain the result. Extension of the injection site into the lateral parts of lobule V is also not a likely explanation, since there is evidence of spread to lobule V in only one case. It thus appears that cells in the ventrolateral outgrowth and the dorsal cap project to crus I in the cat, although the exact site of their termination within crus I cannot be specified. Courville and Faraco-Cantin (in press) have also suggested the possibility of such a projection, based on autoradiographic studies in the cat. There may also be a projection from the dorsal cap and ventrolateral outgrowth to crus II, in light of the findings in Kotchabhakdi et al's (1978) case 666.

In the cat the dorsal and ventral paraflocculi both receive afferents from the ventrolateral part of the rostral medial accessory olive and from the caudolateral bend region of the principal olive, as indicated by HRP studies by Walberg et al. (1979) and in the author's own material. This is somewhat different from the situation in the rabbit (Hoddevik and Brodal A 1977), where olivary afferents to the dorsal paraflocculus arise from the rostrolateral part of the medial accessory olive, while afferents to the ventral paraflocculus take origin from cells in the lateral part of the ventral lamella of the principal olive. In the opossum, segregated olivary cell groups have also been shown to provide input to two distinct areas of the paraflocculus (Linauts and Martin 1978). The lateral paraflocculus receives afferents from the rostral part of the medial accessory olive, while the medial paraflocculus (thought to be comparable to the ventral paraflocculus of other species) gets input from the ventral lamella of the principal olive. Although it appears that there is no comparable segregation of olivary areas providing input to the dorsal and ventral paraflocculi in the cat, two recent autoradiographic studies suggest that there is a clear zonal organization of the olivocerebellar projection to the paraflocculus in this species. According to Groenewegen et al. (1979) and Kawamura and Hashikawa (1979), the paraflocculus can be subdivided into two zones: an inner zone, made up of the adjoining areas of the dorsal and ventral paraflocculi, which receives afferents from the medial accessory olive; and an outer zone, made up of the regions of the dorsal and ventral paraflocculi which lie adjacent to crus II and the flocculus, respectively, which receives afferents from the principal olive.

Reports concerning olivary projections to the flocculus in the rabbit and cat (Alley et al. 1975; Hoddevik and Brodal A 1977; Yamamoto 1979; Walberg et al. 1979; present study) are in agreement that olivary afferents to the flocculus derive mainly from the dorsal cap and adjoining part of the ventrolateral outgrowth and from the rostral tip of the medial accessory olive. According to Walberg et al. (1979) the caudolateral bend region of the principal olive also contains some cells which project to the flocculus.

Recent HRP studies of afferents to the inferior olivary complex in the cat show that olivary inputs arise from a large number of brain stem nuclei including the nucleus subparafascicularis, the fields of Forel, the pretectum, the lateral and middle tegmentum, the nucleus of Darkschewitsch, the ventral periaqueductal gray (Bishop et al. 1976b), the interstitial nucleus of Cajal, the cuneate and gracile nuclei (Bishop et al. 1976b; Buisseret-Delmas and Batini 1978), the deep layers of the tectum (Bishop et al. 1976b; Weber 1978), the medial and descending vestibular nuclei, the supravestibular nucleus, the nucleus praepositus hypoglossi, and vestibular cell groups x, g (Saint-Cyr and Courville 1979), and z (Buisseret-Delmas and Batini 1978; Saint-Cyr and Courville 1979). These studies also show that parts of the inferior olivary complex receive input from layer V of the primary and supplementary motor cortices and from a few neurons in the somatosensory cortex (Bishop et al. 1976a,b), lamina VI of the cervical spinal cord (Buisseret-Delmas and Batini 1978) and the dentate and interposed nuclei (Bishop et al. 1976b; Tolbert et al. 1976; Beitz 1976; Buisseret-Delmas and Batini 1978). These HRP studies can provide only limited data on the termination of various inputs within the inferior olive. However, several other studies employing anterograde degeneration or tracer methods provide information on the distribution of various afferents within the olivary complex.

The deep cerebellar nuclei provide afferents to the principal olivary nucleus and the anterior parts of the medial and dorsal accessory olivary nuclei (Berkley and Worden 1978). Fibers passing to the olive apparently arise from neurons in the dentate and interposed nuclei, but not from the fastigial nucleus (Graybiel et al. 1973; Tolbert et al. 1976; Beitz 1976; Buisseret-Delmas and Batini 1978). The projection to the inferior olive, which arises entirely from small neurons in these nuclei, mostly in their ventral parts (Tolbert et al. 1976; Buisseret-Delmas and Batini 1978), is topographically organized. The dentate nucleus sends fibers to the principal olive, with ventral portions of the dentate supplying the ventral lamella of the principal olive and the dorsal part of the dentate supplying the dorsal lamella (Tolbert et al. 1976; Beitz 1976). The anterior interposed nucleus sends fibers to the dorsal accessory olive, while afferents to the medial accessory olive take origin from the posterior interposed nucleus (Tolbert et al. 1976). Neurons in the ventral part of the posterior interposed nucleus may also send a minor projection to the ventrolateral outgrowth and the dorsal cap (Tolbert et al. 1976; Buisseret-Delmas and Batini 1978). Comparison of the olivary areas receiving input from the deep nuclei with the mapping of the olivocerebellar projection suggests that it is primarily the olivary areas which project to the hemispheral cortex and lateral parts of the anterior lobe which receive input from the deep nuclei. Since the dentate and interposed nuclei receive input from these same cerebellar cortical areas it would appear that the cerebelloolivary projection mediates an indirect feedback loop from the cerebellar cortex. Other studies show that the projections from the deep nuclei to the olivary nucleus are reciprocated by a similarly organized projection from the inferior olive to the deep nuclei (Beitz 1976), giving evidence that the cerebello-olivary projections may be involved in both direct and indirect feedback loops.

Proprioceptive and tactile information is apparently conveyed to the inferior olive by the motor cortex, the dorsal column nuclei, the pars caudalis of the spinal trigeminal nucleus, the lateral cervical nucleus, the spinal cord, and the vestibular nuclei. Afferents from the spinal cord and the dorsal column nuclei terminate in overlapping areas in the lateral two-thirds of the dorsal accessory olive and in a caudal region of the medial accessory olive (Brodal A et al. 1950; Boesten and Voogd 1975; Groenewegen et al. 1975; Berkley and Worden 1978). The projections from the spinal cord and dorsal column nuclei onto the dorsal accessory olive appear to be somatotopically organized, so that hindlimb areas are represented more laterally and forelimb areas are represented more medially (Brodal A et al. 1950; Boesten and Voogd 1975; Groenewegen et al. 1975; Berkley and Hand 1978). The pars caudalis of the spinal trigeminal nucleus projects to the medial zone of the dorsal accessory olive and the adjoining part of the ventral lamella of the principal olive, rostrally. In the medial accessory olive the trigeminal projection zone mostly overlaps with that of the dorsal column nuclei (Berkley and Hand 1978). The inputs terminating in the dorsal accessory olive are conveyed to the intermediate part of the anterior lobe, the medial half of crus I, crus II, and lobulus simplex, and to the medial third of the paramedian lobule. The somatotopic organization evident in the terminal distribution of spinal, trigeminal, and dorsal column afferents in the dorsal accessory olive is apparently preserved by the projection to the anterior lobe and paramedian lobule, so that olivary forelimb areas project on cortical forelimb areas, olivary hindlimb areas project on cortical hindlimb areas, and so on. The region in the caudal medial accessory olive which receives input from the spinal cord, dorsal column nuclei, and trigeminal nucleus appears to project to widespread areas of the vermis, including the vermis proper of the anterior lobe, the vermal visual area, and the lateral part of the uvula.

The olivary projection area of the lateral cervical nucleus (Berkley and Worden 1978) appears to overlap almost entirely with that of the spinal trigeminal nucleus and will not be described here separately.

The primary motor cortex sends fibers mainly to the medial parts of the dorsal and medial accessory olives throughout their rostrocaudal extent and to the ventral lamella of the principal olive, mostly medially, and to a lesser extent to the dorsomedial cell column, the ventrolateral outgrowth, and the dorsal cap (Sousa-Pinto and Brodal A 1969; Berkley and Worden 1978). These regions have been shown to project mainly to parts of the anterior lobe, lobule VII, and the hemispheral cortex, and to a lesser extent on the flocculonodular lobe and uvula. Although the projection to the olive is somatotopically organized (Sousa-Pinto and Brodal A 1969), an attempt to relate the somatotopic organization of the cerebroolivary inputs to the pattern of the olivocerebellar projection as revealed by HRP studies does not reveal any clear correlation with the physiologically demonstrated somatotopic organization of the cerebellar cortex.

A recent HRP study by Saint-Cyr and Courville (1979) shows that the inferior olive receives afferents from the medial and descending vestibular nuclei, the supravestibular nucleus, cell groups x, g, and z, and the nucleus praepositus hypoglossi. Autoradiographic experiments indicate that the vestibular nuclei project bilaterally to the dorsomedial cell column and ipsilaterally to the nucleus β. These regions of the inferior olive in turn project to the medial zone of the uvula, and possibly to the nodulus.

Visual inputs may also be relayed to the cerebellar cortex by the inferior olive. Afferents from the deep layers of the superior colliculus terminate in a rather circumscribed medial area of the caudal medial accessory olive in the cat (Graham 1977; Weber et al. 1978). This area in turn projects preferentially to lobule VII, which is part of the so-called vermal visual area. There is also evidence that a smaller contingent of tectal afferents terminate in the nucleus β (Graham 1977). If this is the case, tectal inputs may also be relayed to the uvula.

Anterograde degeneration studies indicate that fibers originating in the nucleus of Darkschewitsch and the adjacent ventral periaqueductal gray supply parts of the rostral principal olivary nucleus and rostromedial medial accessory olive, including the dorsomedial cell column (Walberg 1974). These olivary regions in turn project to the uvula and parts of the hemispheral cortex. The mesencephalic reticular formation projects to these same olivary areas and, in addition, to the nucleus β and the adjacent region of the caudal medial accessory olive, regions which have strong connections with the vermal visual area and uvula.

Recent anatomic studies in the rabbit have provided evidence of another visual input to the inferior olive. According to Mizuno et al. (1973) after lesions of the pretectum terminal degeneration is found mostly in the dorsal cap of the inferior olivary complex and to a lesser extent in the β nucleus. These olivary areas project mainly to the flocculonodular lobe and the uvula in rabbits. Horseradish peroxidase studies, in which the enzyme is injected in and around the dorsal cap, confirm that projections to this olivary region arise from neurons in the nucleus of the optic tract and the dorsal and lateral terminal nuclei of the accessory optic tract in the rabbit (Takeda and Maekawa 1976). A few neurons projecting to the region of the dorsal cap can also be found in the nuclei of Darkschewitsch and Cajal, the nucleus of the posterior commissure, the dorsal part of the nucleus pretectalis anterior, and the mesecephalic reticular formation. Physiologic studies suggest that the nucleus of the optic tract also projects monosynaptically to the inferior olive in cats, although the exact site of termination of this projection has not been specified (Hoffman et al. 1976).

4.9 Lateral Reticular Nucleus

Brodal's (1943) retrograde degeneration studies of the projection from the lateral reticular nucleus to the cerebellum suggested that practically all neurons in the lateral reticular nucleus send fibers to the ipsilateral cerebellum, and that the fibers were distributed to all parts of the cerebellar cortex. Although the projection appeared to be somewhat diffusely organized, there also seemed to be a degree of localization present. The parvicellular part of the nucleus was thought to project on the vermis, while the magnocellular part appeared to send fibers to the paraflocculus, ansoparamedian lobule, and the lateral parts of the anterior lobe. The small subtrigeminal nucleus was supposed to project to the flocculonodular lobe.

Later anterograde fiber tracing studies using degeneration methods and the autoradiographic technique showed that fibers from the lateral reticular nucleus do not terminate throughout the whole cerebellar cortex, but in fairly well localized projection zones within the cerebellar cortex. Autoradiographic studies by Künzle (1975) demonstrated that fibers from the lateral reticular nucleus terminate bilaterally (with ipsilateral predominance) in the anterior lobe and rostral part of lobule VI, and in the caudal part of lobule VII and lobule VIII. In the hemisphere, fibers from the lateral reticular nucleus reach the paramedian lobule, lobulus simplex, and medial parts of crus I and crus II. The projection to the hemisphere is bilateral with ipsilateral preponderance. The caudal part of lobule VI, the rostral part of lobule VII, the uvula, the nodulus, the lateral parts of crus I and II, the paralocculus, and the flocculus all did not appear to receive a projection from the lateral reticular nucleus. Fairly similar findings were obtained by Matsushita and Ikeda (1976) using anterograde degeneration methods to study the projection from the lateral reticular nucleus onto the cerebellum, except that the projections to lobule VIII and the paramedian lobule appeared to be more restricted in their distribution and the hemispheral projection appeared to be entirely ipsilateral.

Recent studies by Brodal P (1975) and Dietrichs and Walberg (1979), using the HRP method, show that the entire cerebellar cortex receives a projection from the lateral reticular nucleus. The projection is bilateral with strong ipsilateral predominance. The strongest connections are with the anterior lobe, the paramedian lobule and lobule VIIIB. Both the main part of the lateral reticular nucleus (magno- and parvicellular subdivisions) and the subtrigeminal part project to these areas. The projections to the anterior lobe and paramedian lobule evidence a crude topographical organization, so that the lateral part of the main nucleus projects to the rostral part of the anterior lobe and the caudal part of the paramedian lobule (hindlimb areas) while the medial part of the nucleus projects to the caudal parts of the anterior lobe and the rostral parts of the paramedian lobule (forelimb areas). This apparently reflects the somatotopy of the spinal cord input to the lateral reticular nucleus (Brodal P 1975), discussed below. The magnocellular and subtrigeminal portions of the lateral reticular nucleus project to a lesser extent on the other parts of the vermis, crus I, and the flocculus. The projection to crus II arises only from the main body of the lateral reticular nucleus, while the projection to the paraflocculus derives only from the subtrigeminal part (Dietrichs and Walberg 1979).

In contrast to the results of Dietrichs and Walberg (1979), Batini et al. (1978) failed to obtain cell-labelling in the lateral reticular nucleus in cases with injection sites restricted to lobule VII. Their findings after injections in lobule VI appear to be in close agreement with those of Dietrichs and Walberg (1979), however. According to Künzle (1976), the lateral reticular nucleus projects only to the rostral part of lobule VI and the caudal part of lobule VII, so the negative findings of Batini et al. (1978) in cases with injection sites centered in lobule VII may be due to the particular placement of the injection site within the lobule.

The results of the author's investigation are largely in agreement with those of Dietrichs and Walberg (1979). In the present series, however, retrogradely labelled neurons were present only in the subtrigeminal part of the lateral reticular nucleus in the case with an injection site centered in the caudal uvula and nodulus. No labelled cells could be found in any portion of the lateral reticular nucleus in any of the cases with injection sites in the paraflocculus or flocculus. These negative findings have no obvious explanation. It is worthy of note, however, that Alley et al. (1975) reported cell-labelling only in the subtrigeminal part after HRP injections in the flocculus and nodulus of the rabbit. Furthermore, Künzle (1975) saw no evidence of a projection from the main part of the lateral reticular nucleus to the flocculus or nodulus using autoradiographic tracing methods in the cat. (The subtrigeminal part was not investigated.)

Also in contrast to the findings of Dietrichs and Walberg (1979), cell-labelling was almost entirely restricted to the subtrigeminal part of the nucleus in a case with an injection site in crus II. This result must be interpreted with caution, however, since the injection site has spread to adjoining parts of the paraflocculus and paramedian lobule, which receive afferents from the subtrigeminal part of the lateral reticular nucleus according to Dietrichs and Walberg (1979).

The lateral reticular nucleus receives input from a variety of sources, including the spinal cord (Brodal A 1949; Morin et al. 1966; Künzle 1973; Corvaja et al. 1977), the cerebral cortex (Kuypers 1958; Brodal P et al. 1967), the red nucleus (Walberg 1958; Courville 1966; Edwards 1972; Corvaja et al. 1977), the fastigial nucleus (Walberg and Pompeiano 1960; Corvaja et al. 1977), the superior colliculus (Kawamura et al. 1974), and certain of the vestibular nuclei (Ladpli and Brodal A 1968; but see also Corvaja et al. 1977). There is extensive overlap of afferents from different sources within the nucleus. Inputs from the fastigial nuclei, vestibular nuclei, and sensorimotor cortex terminate more medially, while spinoreticular and rubroreticular fibers and more in the lateral part of the nucleus. Rubroreticular fibers terminate rostrally, while the spinoreticular fibers terminate caudally in the lateral part, however. The small subtrigeminal part receives input from the contralateral fastigial nucleus and red nucleus. Despite the extensive overlap of afferents from many sources the lateral reticular nucleus displays a crude somatotopical organization, evident in the pattern of the rubroreticular (Corvaja et al. 1977) and spinoreticular projections (Brodal A 1949; Morin et al. 1966; Künzle 1973; Corvaja et al. 1977). Contrary to previous studies, the results of recent investigations using the HRP method indicate that the somatotopy of the lateral reticular nucleus is preserved in the organization of the reticulocerebellar projection (Brodal P 1975; Dietrichs and Walberg 1979).

4.10 Paramedian Reticular Nucleus and Nucleus Interfasciculares Hypoglossi

Brodal A (1953) and Brodal A and Torvik (1954) described a projection to the cerebellum arising from three cell groups comprising the paramedian reticular nucleus (the dorsal, ventral, and accessory groups) and from scattered nerve cells along the hypoglossal root fibers comprising the nucleus interfasciculares hypoglossi. The majority of the neurons in these reticular cell groups appeared to send fibers to the cerebellum. According to Brodal and Torvik (1954) all three groups of the paramedian reticular nucleus and the nucleus interfasciculares hypoglossi send fibers to the anterior lobe, to lobules VIII and IX, and probably to parts of lobule VII. There was no evidence of any connection with the hemispheral cortex.

The cerebellar connections of the paramedian reticular nucleus of the cat have been reexamined recently by Somana and Walberg (1978a). Based on experiments using the HRP method, they concluded that fibers from the paramedian reticular nucleus terminate over most of the cerebellar cortex. The strongest projection is to the anterior lobe and to the vermis of the posterior lobe. Fewer cells in the paramedian reticular nuclei appear to project to crus I and II, lobulus simplex, and the flocculus. The paramedian lobule and paraflocculus receive only a weak projection. Vermal lobules VIIB and VIIIA, the middle folia of the paramedian lobule, the caudal part of the paraflocculus, and most of the flocculus could not be shown to receive a projection.

In agreement with the earlier retrograde degeneration studies, all three subgroups of the paramedian reticular nucleus (dorsal, ventral, and accessory groups) project to the cerebellar cortex. (Findings concerning the nucleus interfasciculares hypoglossi are not mentioned.) The subgroups project in different proportions to the various parts of the cerebellar cortex, however.

The lingula, lobules VI, VIIA, and VIIIB, and the nodulus receive afferents from all three subgroups. (Compatible findings have been reported by Batini et al. 1978. Injections of HRP into lobule VI elicited cell-labelling in all subgroups of the paramedian reticular nucleus and the nucleus interfasciculares hypoglossi. Injections confined to lobule VII produced negative results, but Batini et al. (1978) did not state which parts of lobule VII were injected.) The uvula and the rostral part of the flocculus receive input only from the accessory and dorsal groups, while lobules II-V, crus I, and lobulus simplex are projected on by the dorsal and ventral groups only. The dorsal group is the major source of paramedian reticular afferents to crus II, the paraflocculus, and the paramedian lobule.

The author's investigations are largely in agreement with those of Somana and Walberg (1978a) concerning the overall distribution of paramedian reticular afferents in the cerebellar cortex. The author's experiments suggest that the strongest projection is to the anterior lobe. (Neurons in the nucleus interfasciculares hypoglossi also project to the anterior lobe.) The posterior lobe vermis appeared to receive a moderate projection but, in contrast to Somana and Walberg's (1978a) findings, the case with an injection site involving most of lobule IX yielded negative results, and in the case with an injection site centered in the caudal part of the uvula and the nodulus HRP-positive neurons are found only in the accessory group. The negative findings in the author's experiments could be due to different placement of the injection site or different amounts of HRP solution injected. The author's material gives evidence of a relatively weak projection to crus I and II and lobulus simplex. The findings in the present study suggest, however, that crus I and lobulus simplex receive input from all three subgroups of the paramedian reticular nucleus, while afferents to crus II arise only from the nucleus interfasciculares hypoglossi.

Although Somana and Walberg (1978a), found evidence of only a weak projection from the dorsal group to the paramedian lobule, the author's investigation suggest that the paramedian lobule has moderately strong connections with the dorsal and accessory groups and the nucleus interfasciculares hypoglossi. The HRP injection site in the author's case CHRCB 10 is larger than any of Somana and Walberg's (1978a), involving most of the paramedian lobule. With this larger injection site, more diffusely branched or sparse inputs to the paramedian lobule might be expected to transport HRP retrogradely, in sufficient amounts to be visible at the light microscopic level.

The author's investigations produced no evidence of a projection to the paraflocculus. This is not surprising, however, as the injection sites in two cases are in the caudal parts of the dorsal and ventral paraflocculi, which do not receive a projection, according to Somana and Walberg (1978a). The injection site in the third case is at the rostral pole of the dorsal paraflocculus, in the junctional area between the dorsal and ventral paraflocculi, and overlaps only slightly with the area shown by Somana and Walberg (1978a) to receive afferents from the dorsal group of the paramedian reticular nucleus.

Concerning the projection to the flocculus, the author's findings are completely compatible with those of Somana and Walberg (1978a).

Afferents to the paramedian reticular nucleus arise from the spinal cord (Brodal A and Gogstad 1957), the fastigial nucleus (Thomas et al. 1956; Walberg et al. 1962), the vestibular nuclei (Ladpli and Brodal A 1968), the superior colliculus (Kawamura et al. 1974; Graham 1977), and the cerebral cortex, especially the primary motor areas (Brodal A and Gogstad 1957; Sousa-Pinto 1970). The dorsal and accessory groups receive input from the face area of motor cortex, while the rest of the senso-rimotor cortex provides a scanty projection to the ventral group.

4.11. External Cuneate Nucleus

The cuneocerebellar projection in the cat has previously been studied by anterograde and retrograde methods. Retrograde degeneration experiments by Brodal A (1941) indicated that the external cuneate nucleus sends fibers only to the anterior lobe and to lobules VIII-IX of the posterior vermis. Grant (1962) suggested a rather different distribution within the cerebellar cortex for the cuneocerebellar fibers, using antero-grade degeneration techniques, however. The results of these studies indicated that the projections from the external cuneate nucleus terminate ipsilaterally in the posterior part of the anterior lobe and the anterior part of lobule VI, in lobule VIII, and in the paramedian lobule.

Reinvestigation of the projection from the external cuneate nucleus to the cere-bellum by the HRP method yields results which largely support Grant's delineation of the terminal fields of afferents from the external cuneate nucleus. According to Rinvik and Walberg (1975) the external cuneate nucleus projects ipsilaterally to the intermediate (and possibly lateral) part of lobule V, the paramedian lobule, and possibly to lobule VIIIB. Retrogradely labelled neurons were found in the external cuneate nucleus only in cases where the injection site extended to, or was centered in, the basal folia of these lobules.

Liu (1956) has indicated that dorsal root afferents terminate in the external cuneate nucleus in a complex somatotopical pattern. Rinvik and Walberg's (1975) findings suggest that this somatotopy is preserved in the organization of the cuneo-cerebellar projection. Thus, the caudal pole of the external cuneate nucleus, which receives afferents from the first cervical root, sends fibers to the caudal part of lobule V and the rostral tip of the paramedian lobule. More rostral parts of the external cuneate nucleus, which receive afferents from the lower cervical (and thoracic) roots, project to more rostral parts of lobule V and more caudal parts of the paramedian lobule.

The author's findings are in excellent agreement with those of Rinvik and Walberg (1975), except concerning the laterality of the projection to the paramedian lobule. (No statement can be made concerning the laterality of the projection to lobule V based on the author's material. as the injection sites involving this lobule are all, at least to some extent, bilateral.) According to Rinvik and Walberg (1975) the fibers passing to the paramedian lobule are strictly ipsilateral. The present study and that of Cheek et al. (1975, see their Fig. 4) show that there is also a small ipsilateral projection.

In one of the author's cases with an injection site centered in the flocculus (not investigated by Rinvik and Walberg 1975) a few HRP-positive neurons are present in the caudoventral part of the external cuneate nucleus. The findings in this case are questionable, however, since the injection site extends into the central white matter and brachium pontis.

4.12 Main Cuneate Nucleus

The existence of a projection from the rostral main cuneate nucleus to the paramedian lobule was initially demonstrated by physiologic methods (Cooke et al. 1971). However, since neurons in the cuneate nuclei do not show retrograde changes following cerebellar lesions (Brodal A 1941) confirmatory anatomic evidence for this projection was lacking until quite recently. The advent of the HRP technique permitted the first anatomic demonstration of a cerebellar projection from the main cuneate nucleus (Cheek et al. 1975; Rinvik and Walberg 1975). These studies showed that neurons in the rostral part of the cuneate nucleus send fibers to the intermediate part of lobule V and the paramedian lobule, mainly ipsilaterally (Rinvik and Walberg 1975), with a minor contralateral component to the projection (Cheek et al. 1975). There was also evidence of a small projection to lobule VIIIB arising from the border zone between the external cuneate and main cuneate nuclei. Although the dorsal root afferents to the cuneate nucleus have been shown to terminate somatotopically within the cuneate nucleus (Rustioni and Macchi 1968) these investigations do not show whether or not the somatotopical pattern is preserved by the organization of the projection to the cerebellum.

The author's own findings are fully in agreement with those of Rinvik and Walberg (1975) and Cheek et al. (1975) and do not add any new information.

4.13 Gracile Nucleus

As was the case for the cuneocerebellar projection, physiologic experiments initially demonstrated the existence of a fiber pathway from the gracile nucleus to the anterior lobe of the cerebellum (Gordon and Horrobin 1967). Recent HRP experiments by Rinvik and Walberg (1975) and Cheek et al. (1975) provided the first anatomic demonstration of this projection. Both groups indicate that fibers from the rostral part of the gracile nuclei terminate in the anterior lobe. However, according to Rinvik and Walberg (1975) the terminal field is in lobules I and II, while Cheek et al. (1975) obtained evidence for a terminal field in the intermediate zone and vermis proper of lobules IV-V.

In contrast, the author's experiments provide evidence of a slightly more widespread distribution of afferents from the gracile nucleus within the cerebellar cortex. Horseradish peroxidase injections centered in lobules IV-V, lobules V-VII, and lobules VIII-IX elicited cell-labelling in the rostral part of the gracile nucleus (lobules I-III were not investigated). Injections in lobule IX or lobules IX-X did not elicit cell-labelling in the gracile nuclei, so gracile afferents probably terminate in lobule VIII, but not in the uvula. Negative results were obtained in all of the cases with injection sites confined to parts of the posterior lobe hemispheral cortex.

4.14 Nucleus of the Solitary Tract

A projection from the nucleus of the solitary tract to the cerebellar vermis and the flocculus in the cat has recently been described by Somana and Walberg (1979a). The afferents from the nucleus of the solitary tract apparently terminate sparsely in the cerebellar cortex, as cell-labelling in this nucleus is elicited inconsistently, and usually only after fairly large HRP injections. The projection does, however, appear to be topographically organized: the rostral part of the nucleus of the solitary tract sends fibers to the posterior vermis, while the caudal part of the nucleus innervates the anterior lobe vermis.

Only a few references to cerebellar projections arising from the nucleus of the solitary tract can be found in the literature. Batini et al. (1978) found retrogradely labelled neurons in the nucleus of the solitary tract in only one out of seven cases in which HRP was injected into vermal lobule VI of the cat. Chan Palay (1977) also reported cell-labelling in the nucleus of the solitary tract in the monkey in a case with an HRP injection into lobule IVA.

The author's experiments provide no evidence of a cerebellar projection arising from the nucleus solitarius. The nucleus contained no clearly labelled neurons even in cases with large injection sites involving most of the cerebellar cortex on one side. Occasionally, neurons containing sparsely distributed granules suggestive of endogenous peroxidase activity were observed in the nucleus of the solitary tract, however.

4.15 Other Brain Stem Nuclei Projecting to the Cerebellar Cortex

Recent experiments in the cat and monkey show that variable numbers of neurons in the cranial nerve motor nuclei send fibers to the cerebellar cortex. Kotchabhakdi and Walberg (1977) found a small number of HRP-positive neurons in the hypoglossal nucleus, the nucleus ambiguus, the motor nucleus of the trigeminal nerve, and the oculomotor, trochlear, and abducens nuclei after HRP injections in the cortex of the anterior and flocculonodular lobes of the cat and monkey. According to these authors, somewhat greater numbers of cranial nerve motor neurons are labelled after HRP injections in the monkey than in the cat. Chan-Palay (1977) has also reported cell-labelling in the dorsal motor nucleus of the vagus nerve and in the oculomotor nucleus following HRP injections into lobule IVA in the monkey.

No cell-labelling was detected in the cranial nerve motor nuclei in the author's experiments in any case of the superficial series, with one exception. In the case with a superficial floccular injection site, as in the case with a nonsuperficial floccular injection site, HRP-positive neurons are present bilaterally in the abducens nuclei and the hypoglossal nuclei. In one case the motor trigeminal nucleus contains a few labelled neurons, bilaterally, but this is likely to be due to encroachment of the injection site on the trigeminal nerve (see comments in Sect. 4.5). The absence of cell-labelling in the author's material in several of the nuclei which have been shown to project to the cerebellar cortex could be attributable to several factors. First, according to Kotchabhakdi and Walberg (1977) only a small number of neurons in cranial nerve motor nuclei are labelled by injections confined to superficial cerebellar cortex in the cat. If labelling occurs in only one or two cells of a nucleus it could easily be missed, as not every section was examined in most cases. Second, the nucleus ambiguus and motor trigeminal nucleus were labelled only following injections involving the ventralmost folia of the anterior lobe (lobules I and II) in Kotchabhakdi and Walberg's (1977) study. In the present study these lobules were never successfully injected in cases of the superficial series. Third, absence of cell-labelling in the oculomotor and trochlear nuclei after injections in the flocculonodular lobe could be due to procedural differences, or to the particular location of the injection sites in the present series.

Several recent HRP studies provide evidence of a projection from the parabrachial nuclei to parts of the cerebellar cortex. According to Somana and Walberg (1979b) neurons in the parabrachial nuclei send fibers to nearly the entire cerebellar vermis, as well as to parts of the ventral paraflocculus and paramedian lobule in the cat. This confirms an earlier report by Batini et al. (1978) of a projection from the parabrachial nuclei to lobule VII of the vermis in the same species. A similar pathway is apparently present in the monkey, as Chan-Palay (1977) found labelled cells in the medial and lateral parabrachial nuclei after HRP injections into lobule IVA in the monkey.

Scattered neurons embedded within the restiform body in a cell group known as the paratrigeminal nucleus (Chan-Palay 1978) also project to parts of lobules I-V and VIIIB-IX and to the paramedian lobule (Somana and Walberg, 1979c). The close proximity of the neurons in this cell group to vestibular cell group x and the external cuneate nucleus, which project on the cerebellar cortex in a similar pattern, raises the question, however, as to whether these scattered cells constitute an independent cell group or are merely displaced neurons belonging to adjacent precerebellar nuclei.

A disynaptic pathway from the retina to the cerebellar cortex has recently been demonstrated in several species. Retinal inputs are apparently conveyed to the cerebellum by way of the medial terminal nucleus of the accessory optic tract (or its homologs) in the rabbit (Winfield et al. 1978), pigeon (Brauth and Karten 1977; Brecha et al. 1977), turtle (Reiner and Karten 1978), and fish (Finger and Karten 1977). To date, there do not appear to be any reports of a comparable pathway in cats. The author's material provides no information on this issue, as sections were not routinely cut rostral to the griseum pontis. Whether or not a disynaptic pathway for retinal input to the cerebellar cortex exists in cat remains, therefore, an unanswered question.

5 General Discussion of the Afferent Organization of the Cerebellar Cortex of the Cat

5.1 Overlap and Segregation of Brain Stem Areas Projecting to Different Cerebellar Lobules

Many of the recent studies using retrograde and anterograde axoplasmic transport methods (reviewed above) have confirmed the existence of projections from the brain stem nuclei to the cerebellar cortex which had previously been demonstrated by classic anatomic methods. Still other investigations have provided evidence of cerebellar cortical afferents arising from nuclei which could not be shown to project to the cerebellar cortex by degeneration methods. Even within the nuclei which had previously been known to send fibers to the cerebellar cortex, such as the pons and the inferior olive, the distribution of neurons projecting to particular cerebellar cortical lobules appears more widespread than was previously believed. This raises a question as to whether particular regions of a nucleus can project to more than one cerebellar lobule.

The answer to this question appears to be affirmative. For example, in the pons, the rostral part of the peduncular nucleus sends fibers to both the paramedian lobule and uvula (Brodal A and Hoddevik 1978); the dorsolateral nucleus provides afferents to vermal lobule VII (Hoddevik et al. 1977) and to the paramedian lobule (Hoddevik 1975). A similar situation obtains in many of the other brain stem precerebellar nuclei. The central area of the superior vestibular nucleus sends fibers to both the flocculus and the nodulus (Kotchabhakdi and Walberg 1978b). The caudal part of the medial accessory olive projects to nearly all of the vermis proper in the cat (Brodal A and Walberg 1977a; Hoddevik et al. 1976). The magnocellular part of the lateral reticular nucleus maintains connections with practically all of the cerebellar cortex (Dietrichs and Walberg 1979), as does the dorsal group of the paramedian reticular nucleus (Somana and Walberg 1978a). There are many other examples of restricted regions of brain stem precerebellar nuclei which appear to provide afferents to more than one cerebellar lobule, but they are too numerous to list here.

Even if small parts of precerebellar nuclei project to several cerebellar lobules, it could be the case that separate populations of neurons within a restricted region send fibers to different cerebellar lobules. This raises a second question: Do single neurons have axon collaterals terminating in more than one cerebellar lobule? No satisfactory answer to this question appears to be available at present, as studies using the HRP method alone cannot provide conclusive information on this point. However, indirect evidence from these studies supports the notion that at least some neurons in the external cuneate nucleus may send axon collaterals both to lobule V and to the paramedian lobule. When HRP is injected into both lobule V and the paramedian lobule, the number of retrogradely labelled neurons is far greater than the sum of labelled cells in the nucleus in two cases with separate injections in lobule V and the paramedian lobule (Rinvik and Walberg 1975). (Apparently, some neurons projecting to a cerebellar lobule may not be visible with the HRP technique unless all of their axonal ramifications are exposed to the enzyme.) Axon collateralization might explain the failure of retrograde degeneration methods to demonstrate the projection from the external cuneate nucleus to the paramedian lobule (Brodal A 1941). Electrophysiologic investigations also suggest that single olivary neurons may send axon collaterals to several folia within a sagittally oriented olivary projection zone (Armstrong et al. 1973). Experiments using two retrograde markers injected into separate cerebellar cortical areas could be useful in providing definitive anatomic evidence of axon collateralization, and in determining whether this is a general characteristic of neurons in precerebellar nuclei (see, for example, Geisert 1976, Weisberg and Metz 1976; Steward et al. 1977; van der Kooy et al. 1978; Hayes and Rustioni 1979).

5.2 Overlap and Segregation of Afferents from Different Brain Stem Areas Within the Cerebellar Cortex

The multiplicity of brain stem nuclei projecting to the cerebellar cortex and the widespread distribution of afferents from a particular nucleus within the cerebellar cortex suggest that it is almost certainly the case that there is considerable overlap of inputs from different brain stem areas within the cerebellar cortex. Although there does not appear to be any direct anatomic support for this notion, indirect evidence suggests that some of the mossy fiber inputs may converge on small cerebellar cortical regions. In cases with very small HRP injection sites in the cerebellar cortex, cell-labelling is elicited in several longitudinally oriented cell columns in the pontine nuclei (Hoddevik 1975, Hoddevik et al. 1977; Hoddevik and Walberg 1979).

On the other hand, climbing fiber inputs from different parts of the inferior olive may not overlap at all in the cerebellar cortex. Although afferents to the paramedian lobule arise from four separate areas of the inferior olive (Brodal A et al. 1975), analysis of the olivocerebellar projection in cases with very small HRP injection sites in the paramedian lobule shows that the cortex of this lobule is divided into four longitudinal zones whose olivary afferents differ. A similar strict parcellation of zones receiving afferents from particular portions of the olivary complex also appears to obtain for the rest of the cerebellar cortex (discussed in more detail below).

Experiments using two anterograde fiber tracing methods to compare the cerebellar projections from different sources in a single animal could be of considerable value in providing information on the terminal distribution and relationships between afferents from different nuclei in the cerebellar cortex. This technique has already been used successfully to study the terminal distributions of various afferents to the inferior olive (Berkley and Worden 1978; Berkley and Hand 1978). It should be borne in mind, however, that even if mossy fibers from different sources could be shown to terminate in discrete patches in the granule cell layer of the cerebellar cortex, it is unlikely that segregation of inputs could be maintained at the next level of processing, as recent estimates suggest that parallel fibers span a length of 5-7 mm (Brand et al. 1976).

5.3 Sagittal Organization of the Cerebellar Cortex

The notion that the cerebellar cortex could be divided into longitudinally oriented zones was originally proposed on the basis of its afferent (Brodal A 1940) and efferent organization (Jansen and Brodal A 1940). These early experiments showed that afferents from particular subdivisions of the inferior olivary complex terminate in sagittally oriented strips of cerebellar cortex. Moreover, the cerebellar cortical outflow to the deep cerebellar nuclei also appeared to be organized longitudinally, so that the fastigial nucleus received input from the cortex of the vermis proper, while the paravermal cortex projected on the interposed nuclei and the hemispheral cortex projected on the dentate nucleus. Subsequent investigations indicated that some of the details of this schema needed to be modified (Walberg and Jansen 1964; Brodal A and Courville 1973; Courville et al. 1973; Courville and Diakiw 1976; Courville and Faraco-Cantin 1976) but the weight of the evidence generally favors the notion that the cerebellar cortex can be divided into a series of longitudinal zones whose afferent and efferent connections differ.

Voogd (1964) described "raphes" in the central and folial white matter in myelin-stained sections of cat cerebellum. These raphes formed the borders of several longitudinally oriented compartments in the cerebellum (the compartments are actually longitudinally oriented only in the vermis; the situation is more complicated in the hemisphere, where the folia are not arranged in a straight line). According to this scheme, compartment A, on the midline of the vermis proper, projected to the fastigial nucleus. Compartment B, just lateral to that, projected to the lateral vestibular nucleus. Compartment C, the next most lateral, projected to the "pars convexa" of the dentate nucleus and to the anterior interposed nucleus. The corticofugal connections of the lateralmost zone, D, could not be determined. In the hemisphere, three roughly circumferentially arranged zones were described, with compartment B^1 situated more centrally, compartment D^1 peripherally, and compartment C^1 in between. Compartment B^1 appeared to project to the posterior interposed nucleus, while C^1 projected to the pars convexa of the dentate nucleus, and D^1 projected to the pars rotunda of the dentate nucleus. Two recent studies of the cortical projection to the deep cerebellar nuclei and the lateral vestibular nucleus using the HRP method (Courville and Faraco-Cantin 1976, Corvaja and Pompeiano 1979) suggest a zonal subdivision of the cerebellar cortex which is remarkably similar to the schema proposed by Voogd (1964).

Other recent studies suggest that some afferents to the cerebellum terminate in similarly organized longitudinal zones in the cerebellar cortex. Afferents from the inferior olive terminate in narrow bands in the cerebellar cortex, traversing many folia (Voogd 1969; Courville 1975). Detailed analysis of the olivocerebellar projection has indicated that some of Voogd's compartments can be subdivided further. Thus, according to Groenewegen and Voogd (1977) and Groenewegen et al. (1979), zone A, encompassing the medialmost strip of vermis from lobules I-IX, receives afferents from the caudal medial accessory olive. Zone B, which is the next most lateral strip in the vermis of lobules I-VI, is projected on by the caudal dorsal accessory olive. Zones C_{1-3} and D, involving successively more lateral strips of the anterior lobe and hemisphere are projected on by neurons in the rostral dorsal accessory olive (C_1 and C_3), the rostral medial accessory olive (C_2) and the principal olive (D). The lateral parts of the uvula receive input from the dorsomedial cell column and rostral medial accessory olive, while the flocculus and nodulus receive afferents from the ventrolateral outgrowth and dorsal cap. Although there is some disagreement on the details of the olivocerebellar projection, investigations using the HRP technique suggest a basically similar organization of the olivocerebellar projection (Brodal A 1976, Hoddevik et al. 1976; Brodal A and Walberg 1977a,b; Kotchabhakdi et al. 1978).

Of the mossy fiber systems, termination in sagittally oriented bands has been demonstrated only for spinal, external cuneate, and lateral reticular nucleus afferents to the anterior lobe and lobulus simplex of the cat (Voogd 1969; Künzle 1975). It is not clear, therefore, whether the mossy fiber systems follow a general plan of organization similar to the climbing fiber system. That projection onto longitudinal zones may not be a general feature of mossy fiber afferents is suggested by the fact that the pontocerebellar projection to the anterior lobe of the cat does not give evidence of sagittal banding (Voogd 1969). Further research using anterograde tracing methods will be necessary to determined how each of the mossy fiber systems terminates in the cerebellar cortex, and whether different mossy fiber inputs terminate in register or in adjacent patches of cerebellar cortex.

5.4 Afferent Organization of the Cerebellar Cortex in Relation to Concepts of Functional Localization

On the basis of previous studies of its afferent connections, the cerebellar cortex has been subdivided into three major functional areas: the pontocerebellum, consisting of the lateral part of the anterior lobe, the hemispheres, and the midvermal cortex;

the spinocerebellum, which includes the vermis and intermediate zone of the anterior lobe, lobules VIII and IX, and the paramedian lobule; and the vestibulocerebellum, restricted to the flocculonodular lobe, uvula, and ventral paraflocculus. It has been known for some time that there is considerable overlap between these subdivisions (see reviews by Brodal A 1967, 1972). The recent anatomic studies reviewed above suggest that the overlap of the subdivisions is even more extensive than was heretofore believed.

The definition of the vestibulocerebellum must now be expanded to include the entire vermis, as well as the flocculus and nodulus, following the demonstration by Kotchabhakdi and Walberg (1978a, b) that primary and secondary vestibular fibers terminate in all of these areas. Vestibular inputs may also be relayed to the cerebellar cortex by less direct routes. The superior and lateral vestibular nuclei send fibers to parts of the nucleus reticularis tegmenti pontis which project mainly to the anterior lobe and lobules VI-VIII. The medial and descending vestibular nuclei project to the dorsomedial cell column and nucleus β, which in turn project on the uvula (and possibly the nodulus).

The pontine nuclei provide afferents to nearly all of the cerebellar cortex, as was shown originally by Brodal A and Jansen (1946). The nodulus does not appear to receive a pontine input, however, and whether the flocculus receives pontine afferents over its whole extent is still unresolved (see discussion in section on the pontine nuclei, above).

A detailed discussion of inputs to the cerebellar cortex from the spinal cord proper is somewhat outside of the scope of this paper. Suffice it to say that although spinal cord inputs to the cerebellum have recently been shown to arise from areas of the spinal gray which were not previously believed to maintain a cerebellar projection (see, for example, Wiksten 1975; Matsushita and Ikeda 1975; Snyder et al. 1978; Matsushita et al. 1979), there do not appear to be any recent studies in the cat which challege the prior delimitations of the terminal fields of spinocerebellar fibers. Recent studies of cerebellar projections arising from the cuneate, gracile, and external cuneate nuclei suggest that projections from these nuclei terminate entirely within the previously defined spinocerebellum (Rinvik and Walberg 1975; Cheek et al. 1975). However, spinal inputs may also be relayed to the cerebellar cortex by way of the other brain stem precerebellar nuclei, including the pontine nuclei, the inferior olive, the vestibular nuclei, the raphe nuclei, the lateral reticular nucleus, and the paramedian reticular nucleus. Spinopontine fibers terminate mainly in the caudal part of the dorsolateral nucleus, which in turn projects mainly on the anterior lobe and lobules VI-VIII, and to a lesser extent on the uvula, paramedian lobule, and crus I and II. Likewise, regions of the dorsal and medial accessory olives which receive spinal afferents send fibers to the entire vermis and to a slightly more lateral strip of cortex involving the intermediate zone of the anterior lobe and the medial parts of lobulus simplex, crus I and II, and the paramedian lobule. Spinal inputs terminating in the caudal parts of the medial and descending vestibular nuclei and cell group x could be relayed to the entire vermis. The spinal cord also provides afferents to the raphe nuclei magnus and pallidus, the lateral reticular nucleus, and the paramedian reticular nucleus, which project to widespread areas of the cerebellar cortex. Thus, even if afferents from the spinal cord are not relayed directly to cerebellar areas outside of the traditional spinocerebellum, information ascending from spinal levels could still reach other cerebellar cortical areas by way of relays in the brain stem precerebellar nuclei.

Physiologic studies in the cat have also defined areas in the cerebellar cortex which contain somatotopical mappings of the body surface or receive auditory and visual input (Snider and Stowell 1944; Hampson 1949; Snider and Eldred 1951). By and large, the recent studies of brain stem afferents to the cerebellar cortex demonstrate a clearcut anatomic basis for localization of orderly body maps in the anterior lobe and paramedian lobule, and a visual and auditory receiving area in the midvermal cortex. Thus, somatotopically ordered inputs from the somatosensory and motor cortices are relayed to the cortex of the anterior and posterior lobes by way of the pontocerebellar projection. Somatotopically ordered input is also conveyed to these areas by way of rubral and spinal inputs to the lateral reticular nucleus and by way of spinal, trigeminal, and dorsal column nuclear inputs to the inferior olive. Moreover, projections from the trigeminal nuclei terminate in the physiologically identified face areas, while cerebellar afferents from the cuneate nucleus terminate in physiologically identified forelimb areas.

A clearcut basis for localization of an auditory and visual receiving area in lobules VI-VIII is also evident in the organization of the ponto- and olivocerebellar projections. Fibers conveying visual and auditory inputs from the superior and inferior colliculi and the auditory cortex terminate in the dorsolateral pontine nucleus, which in turn projects heavily on the visual and auditory receiving area, especially lobule VII, and to a lesser extent to the hemispheral cortex, anterior lobe and uvula. Visual input also appears to be conveyed to lobule VII from the medialmost part of the caudal medial accessory olive (referred to as subnucleus B by Weber et al. 1978, not to be confused with nucleus β) which receives afferents from the superior colliculus.

It is significant that in the cat (but not in the rat or monkey, see Burne et al. 1978; Brodal P 1978) afferents from visual cortex terminate in a pontine area which is distinct from the region receiving superior colliculus afferents. The results of recent HRP studies suggest that the pontine visual cortex receiving area projects to widespread regions of the cerebellar cortex outside of the vermal visual area, including parts of the hemispheral cortex and the uvula. Thus, visual afferents do not pass exclusively to the vermal visual and auditory receiving area in the cat.

The notion that other cerebellar areas receive visual input is not a new one. Physiologic studies in the rabbit have previously demonstrated that the flocculus, for example, receives visual input by way of both mossy fiber and climbing fiber pathways (Maekawa and Simpson 1973; Maekawa and Takeda 1975). The results of HRP studies suggest that visual inputs may in fact be relayed to some extent to nearly all of the cerebellar cortex of the cat by the pontocerebellar projection, although a major projection passes to the visual and auditory receiving area. This means that visual inputs probably also pass to areas of the cerebellar cortex which have previously been shown to be major receiving areas for somatotopically ordered inputs. Conversely, proprioceptive and tactile impulses may pass to the visual and auditory receiving area by way of relays in the pons, the vestibular nuclei, the perihypoglossal nuclei, the inferior olive, the lateral reticular nucleus, and the main and external cuneate nuclei. This apparently complex pattern of termination of proprioceptive and exteroceptive inputs could provide an anatomic substrate for integration of sensory and motor information in the cerebellar cortex, which would be necessary for planning and execution of skilled movements to auditory and visual targets.

In closing, one final caution should be added. The mere fact that cell-labelling is observed in an area of a nucleus which is known to receive input from a particular source after HRP injections in a circumscribed region of cerebellar cortex does not mean that that nucleus is in fact relaying input from that source to the cerebellar area in question. For example, it has not been proven that the neurons in the dorsolateral pontine nucleus which project to the paramedian lobule are contacted synaptically by fibers from the superior colliculus. Rather, afferents from the superior colliculus may terminate only on neurons which ultimately project to the vermal visual area, which would mean that the apparent diffuseness of projections conveying visual inputs to the cerebellar cortex is purely illusory. This possibility could be investigated experimentally by the use of electron microscopy combined with anterograde and retrograde axoplasmic transport tracer techniques (see Dekker 1979). In such an experiment tritiated amino acids would be injected into a structure known to provide input to a particular precerebellar nucleus and HRP would be injected into an area of the cerebellar cortex known to receive input from that nucleus. After appropriate treatment, areas of the precerebellar nucleus could be examined under the electron microscope to determine whether or not retrogradely labelled neurons are contacted synaptically by radioactively labelled terminals. It is suggested that experiments of this type could be of considerable value in elucidating the synaptic linkages of afferents to the cerebellar cortex.

6 Summary

Recent investigations using retrograde and anterograde tracer methods have confirmed the results of prior studies showing that, in the cat, afferents to the cerebellar cortex arise from the pontine nuclei, the nucleus reticularis tegmenti pontis, the trigeminal nuclei, the vestibular nuclei, the perihypoglossal nuclei, the inferior olive, the lateral reticular nucleus, the paramedian reticular nucleus, and the external cuneate nucleus. In general, the studies using the more sensitive axoplasmic transport methods have provided new details concerning the origin and cortical termination of these afferents to the cerebellar cortex. In addition, the recent investigations have demonstrated that cerebellar cortical afferents arise from several brain stem nuclei which had not previously been shown to project to the cerebellar cortex, including the raphe nuclei, a lateral rhombencephalic tegmental cell group, the locus coeruleus, the main cuneate nucleus, the gracile nucleus, the nucleus of the solitary tract, the parabrachial nuclei, the paratrigeminal nucleus, and certain of the cranial nerve motor nuclei.

Afferents from the *pontine nuclei* terminate throughout the cerebellar cortex, except for the flocculonodular lobe. The nodulus does not appear to receive a projection from the pontine nuclei in the cat; it is unclear whether all parts of the flocculus receive input from the pontine nuclei. The pontocerebellar projection is precisely organized; neurons which send fibers to a particular part of the cerebellar cortex are located in longitudinally oriented columns in the pontine gray.

Clusters of neurons in the main body of the *nucleus reticularis tegmenti pontis* send fibers to the entire cerebellar cortex. Afferents to the flocculus and nodulus arise from circumscribed cell groups in the dorsomedial and dorsolateral parts, respectively, of the main body of the nucleus. Neurons in the processus tegmentosus lateralis project to lobules VI—VIIIA and to a lesser extent to crus I and II and the dorsal paraflocculus.

The *corpus pontobulbare* projects most strongly to the anterior lobe and paramedian lobule, and to a lesser extent to lobules VI—IX, crus I and II, and the paraflocculus.

All parts of the cerebellar cortex, except possibly lobule VI, receive afferents from the *raphe nuclei*. The raphe nuclei pontis and obscurus contain the largest number of neurons projecting to the cerebellar cortex.

Neurons in a *lateral tegmental cell group* at the level of the isthmus project mainly to the middle vermis and crus II, and to a lesser extent to the anterior and posterior vermis, crus I, and the paramedian lobule.

The caudal part of the *locus coeruleus* projects to the entire cerebellar vermis (mainly to its anteriormost and posteriormost parts) and to the flocculus and ventral paraflocculus.

The *principal trigeminal nucleus* and pars oralis and pars interpolaris of the *spinal trigeminal nucleus* project to lobules V—VIIIA, lobulus simplex, the posterior part of crus II, and the dorsal part of the paramedian lobule.

Certain of the *vestibular nuclei* supply afferents to the entire cerebellar vermis and the flocculus. The cerebellar projection arises from the medial, descending, superior, and possibly lateral vestibular nuclei, as well as the supravestibular nucleus and cell groups f, x, and y.

The *perihypoglossal nuclei* project to the entire vermis, the flocculus, and the paraflocculus. The projection to the paraflocculus is quantitatively a very minor one.

The *inferior olive* supplies climbing fibers to the entire cerebellar cortex. Each subdivision of the inferior olivary complex supplies a sagittally oriented strip of cerebellar cortex, which usually spans several cerebellar lobules.

The *lateral reticular nucleus* sends fibers to most of the cerebellar cortex. The strongest projections, arising from all parts of the nucleus, are to the anterior lobe, paramedian lobule, and lobule VIIIB. Other parts of the vermis and the cerebellar hemisphere (possibly only the medial parts) receive a weaker projection.

The *paramedian reticular nucleus* projects to most of the cerebellar cortex. Only lobules VIIB and VIIIA, the middle folia of the paramedian lobule, and the caudal folia of the flocculus and paraflocculus have not been shown to receive input from the paramedian reticular nucleus.

Afferents from the *external cuneate nucleus* reach the intermediate (and possibly lateral) parts of lobule V, the paramedian lobule, and possibly lobule VIIIB. The *main cuneate nucleus* provides a sparse input to these same areas.

The *gracile nucleus* sends a minor projection to the cerebellar vermis. The location of the terminal field of the projection is not agreed upon, but the evidence suggests that gracile nucleus afferents may terminate in both anterior and posterior vermal regions.

The *nucleus of the solitary tract* sends a very sparse projection to the cerebellar vermis and flocculus. The rostral part of the nucleus projects to the posterior vermis, while the caudal part of the nucleus projects to the anterior vermis.

The *parabrachial nuclei* project to most of the vermis and to parts of the paramedian lobule and ventral paraflocculus.

The *paratrigeminal nucleus* sends fibers to parts of lobules I–V, lobules VIII–IX, and the paramedian lobule.

Certain of the *cranial nerve motor nuclei* provide afferents to the cerebellar cortex. These nuclei appear to project mostly to the anteriormost folia of the anterior lobe and to the flocculonodular lobe.

References

Alley K, Baker R, Simpson JI (1975) Afferents to vestibulo-cerebellum and the origin of the visual climbing fibers in the rabbit. Brain Res 98: 582–589

Altman J, Carpenter MB (1961) Fiber projections of the superior colliculus in the cat. J Comp Neural 116: 157–178

Altman JA, Bechterev NN, Radionova EA, Shmigidina GN, Syka J (1976) Electrical responses of the auditory area of the cerebellar cortex to acoustic stimulation. Exp Brain Res 26: 285–298

Andén NE, Fuxe K, Ungerstedt U (1967) Monoamine pathways to the cerebellum and cerebral cortex. Experientia 23: 838–839

Angaut P, Brodal A (1967) The projection of the "vestibulo-cerebellum" onto the vestibular nuclei in the cat. Arch Ital Biol 105: 441–479

Armstrong DM, Harvey RJ, Schild RF (1973) Branching of inferior olivary axons to terminate in different folia, lobules or lobes of the cerebellum. Brain Res 54: 365–371

Baker J, Gibson A, Glickstein M, Stein J (1976) Visual cells in the pontine nuclei of the cat. J Physiol (Lond) 255: 415–433

Baker R, Berthoz A (1975) Is the prepositus hypoglossi nucleus the source of another vestibulo-ocular pathway? Brain Res 86: 121–127

Batini C, Corvisier J, Destombes J, Gioanni H, Everett J (1976) The climbing fibers of the cerebellar cortex, their origin and pathways in cat. Exp Brain Res 26 : 407–422

Batini C, Corvisier J, Hardy O, Jassik-Gerschenfeld D (1977) Perihypoglossal and secondary vestibular projections to lobules VI and VII of the cerebellar cortex: an HRP study. Neuroscience Letters 5 : 111–116

Batini C, Buisseret-Delmas C, Corvisier J, Hardy O, Jassik-Gerschenfeld D (1978) Brain stem nuclei giving fibers to lobules VI and VII of the cerebellar vermis. Brain Res 153: 241–261

Beitz AJ (1976) The topographical organization of the olivo-dentate and dentato-olivary pathways in the cat. Brain Res 115: 311–317

Berkley KJ, Hand PJ (1978) Projections to the inferior olive of the cat. II. Comparisons of input from the gracile, cuneate and spinal trigeminal nuclei. J Comp Neurol 180: 253–264

Berkley KJ, Worden IG (1978) Projections to the inferior olive of the cat. I. Comparisons of input from the dorsal column nuclei, the lateral cervical nucleus, the spino-olivary pathways, the cerebral cortex and the cerebellum. J Comp Neurol 180: 237–252

Berman AL (1968) The brain stem of the cat. A cytoarchitectonic atlas with stereotaxic co-ordinates. University of Wisconsin Press, Madison

Berman N (1977) Connections of the pretectum in the cat. J Comp Neurol 174: 227–254

Bishop GA, McCrea RA, Kitai ST (1976 a) A horseradish peroxidase study of the cortico-olivary projection in the cat. Brain Res 116: 306–311

Bishop GA, McCrea RA, Kitai ST (1976 b) Ascending and descending projections to the inferior olive of the cat. Neuroscience Abstracts 2: 106

Bloom FE, Hoffer BJ, Siggins GR (1971) Studies on norepinephrine-containing afferents to Purkinje cells of rat cerebellum. I. Localization of the fibers and their synapses. Brain Res 25: 501–521

Bobillier P, Seguin S, Petitjean F, Salvert D, Touret M, Jouvet M (1976) The raphe nuclei of the cat brain stem: a topographical atlas of their efferent projections as revealed by autoradiography. Brain Res 113: 449–486

Boesten AJP, Voogd J (1975) Projections of the dorsal column nuclei and the spinal cord on the inferior olive in the cat. J Comp Neurol 161: 215–238

Bowden DM, German DC, Poynter WD (1978) An autoradiographic, semistereotaxic mapping of major projections from locus coeruleus and adjacent nuclei in Macaca Mulatta. Brain Res 145: 257–276

Brand S, Dahl AL, Mugnaini E (1976) The length of parallel fibers in the cat cerebellar cortex. An experimental light and electron microscopic study. Exp Brain Res 26: 39–58

Brauth SE, Karten HJ (1977) Direct accessory optic projections to the vestibulo-cerebellum: a possible channel for oculomotor control systems. Exp Brain Res 28: 73–84

Brecha N, Karten HJ, Hunt S (1977) A visual quickie: a bisynaptic retinal pathway to the vestibulo-cerebellum and oculomotor nuclear complex. Neuroscience Abstracts 3: 554

Brodal A (1940) Experimentelle Untersuchungen über die olivocerebellare Lokalisation. Gesamte Neurol Psychiatr 169: 1–153

Brodal A (1941) Die Verbindungen des Nucleus cuneatus externus mit dem Kleinhirn beim Kaninchen und bei der Katze. Experimentelle Untersuchungen. Gesamte Neurol Psychiat 171: 167–199

Brodal A (1943) The cerebellar connections of the nucleus reticularis lateralis (nucleus funiculi lateralis) in rabbit and cat. Experimental investigations. Acta Psychiat Neurol 18:171–233

Brodal A (1949) Spinal afferents to the lateral reticular nucleus of the medulla oblongata in the cat. An experimental study. J Comp Neurol 91: 259–295

Brodal A (1952) Experimental demonstration of cerebellar connexions from the peri-hypoglossal nuclei (nucleus intercalatus, nucleus praepositus hypoglossi and nucleus of Roller) in the cat. J Anat 86: 110–129

Brodal A (1953) Reticulo-cerebellar connections in the cat. An experimental study. J Comp Neurol 98: 113–154

Brodal A (1954) Afferent cerebellar connections. In: Jansen J, Brodal A (eds) Aspects of cerebellar anatomy. Grundt Tanum, Oslo, pp 82–188

Brodal A (1967) Anatomical studies of cerebellar fibre connections with special reference to problems of functional localization. In: Fox CA, Snider RS (eds) The cerebellum. Progress Brain Research, vol 25. Elsevier, Amsterdam, pp 135–173

Brodal A (1972) Cerebrocerebellar pathways. Anatomical data and some functional implications. Acta Neurol Scand [Suppl] 51: 153–196

Brodal A (1976) The olivocerebellar projection in the cat as studied with the method of retrograde axonal transport of horseradish peroxidase. II. The projection to the uvula. J Comp Neurol 166: 417–426

Brodal A, Angaut P (1967) The termination of spinovestibular fibres in the cat. Brain Res 5: 494–500

Brodal A, Brodal P (1971) The organization of the nucleus reticularis tegmenti pontis in the cat in light of experimental anatomical studies of its cerebral cortical afferents. Exp Brain Res 13: 90–110

Brodal A, Courville J (1973) Cerebellar corticonuclear projection in the cat. Crus II. An experimental study with silver methods. Brain Res 50: 1–23

Brodal A, Gogstadt AC (1957) Afferent connexions of the paramedian reticular nucleus of the medulla oblongata in the cat. An experimental study. Acta Anat (Basel) 30: 133–151

Brodal A, Hoddevik GH (1978) The pontocerebellar projection to the uvula in the cat. Exp Brain Res 32: 105–116

Brodal A, Høivik B (1964) Site and mode of termination of primary vestibulocerebellar fibers in the cat. An experimental study with silver impregnation methods. Arch Ital Biol 102: 1–21

Brodal A, Jansen J (1946) The ponto-cerebellar projection in the rabbit and cat. Experimental investigations. J Comp Neurol 84: 31–118

Brodal A, Saugstad LF (1964) Retrograde cellular changes in the mesencephalic trigeminal nucleus in the cat following cerebellar lesions. Acta Morphol Neerl Scand 6: 147–159

Brodal A, Szikla G (1972) The termination of the brachium conjunctivum descendens in the nucleus reticularis tegmenti pontis. An experimental anatomical study in the cat. Brain Res 39:337–351

Brodal A, Torvik A (1954) Cerebellar projection of paramedian reticular nucleus of medulla oblongata in cat. J Neurophysiol 17: 484–495

Brodal A, Torvik A (1957) Über den Ursprung der sekundären vestibulo-cerebellaren Fasern bei der Katze. Eine experimentell-anatomische Studie. Arch Psychiat Z Gesamte Neurol 195: 550–567

Brodal A, Walberg F (1977a) The olivocerebellar projection in the cat studied with the method of retrograde axonal transport of horseradish peroxidase. IV. The projection to the anterior lobe. J Comp Neurol 172: 85–108

Brodal A, Walberg F (1977b) The olivocerebellar projection in the cat studied with the method of retrograde axonal transport of horseradish peroxidase. VI. The projection onto longitudinal zones of the paramedian lobule. J Comp Neurol 176: 281–294

Brodal A, Walberg F, Blackstad TH (1950) Termination of spinal afferents to inferior olive in cat. J Neurophysiol 13: 431–454

Brodal A, Taber E, Walberg F (1960) The raphe nuclei of the brain stem in the cat. II. Efferent connections. J Comp Neurol 114: 239–260

Brodal A, Walberg F, Taber E (1960) The raphe nuclei of the brain stem in the cat. III. Afferent connections. J Comp Neurol 114. 261–282

Brodal A, Destombes J, Lacerda AM, Angaut P (1972) A cerebellar projection onto the pontine nuclei. An experimental anatomical study in the cat. Exp Brain Res 16:115–139

Brodal A, Lacerda AM, Destombes J, Angaut P (1972) The pattern in the projections of the intracerebellar nuclei onto the nucleus reticularis tegmenti pontis in the cat. An experimental anatomical study. Exp Brain Res 16: 140–160

Brodal A, Walberg F, Hoddevik GH (1975) The olivocerebellar projection in the cat studied with the method of retrograde axonal transport of horseradish peroxidase. J Comp Neurol 164: 449–470

Brodal P (1968a) The corticopontine projection in the cat. I. Demonstration of a somatotopically organized projection from the primary sensorimotor cortex. Exp Brain Res 5: 210–234

Brodal P (1968b) The corticopontine projection in the cat. Demonstration of a somatotopically organized projection from the second somatosensory cortex. Arch Ital Biol 106: 310–332

Brodal P (1971a) The corticopontine projection in the cat. I. The projection from the proreate gyrus. J Comp Neurol 142: 127–140

Brodal P (1971b) The corticopontine projection in the cat. II. The projection from the orbital gyrus. J Comp Neurol 142: 141–152

Brodal P (1972a) The corticopontine projection in the cat. The projection from the auditory cortex. Arch Ital Biol 110: 119–144

Brodal P (1972b) The corticopontine projection from the visual cortex in the cat. I. The total projection and the projection from area 17. Brain Res 39: 297–317

Brodal P (1972c) The corticopontine projection from the visual cortex in the cat. II. The projection from areas 18 and 19. Brain Res 39: 319–335

Brodal P (1975) Demonstration of a somatotopically organized projection onto the paramedian lobule and the anterior lobe from the lateral reticular nucleus: an experimental study with the horseradish peroxidase method. Brain Res 95: 221–239

Brodal P (1978) The corticopontine projection in the rhesus monkey. Origin and principles of organization. Brain 101: 251–283

Brodal P, Walberg F (1977) The pontine projection to the cerebellar anterior lobe. An experimental study in the cat with retrograde axonal transport of horseradish peroxidase. Exp. Brain Res 29: 233–248

Brodal P, Marsala J, Brodal A (1967) The cerebral cortical projection to the lateral reticular nucleus in the cat, with special reference to the sensorimotor cortical areas. Brain Res 6: 252–274

Buisseret-Delmas C, Batini C (1978) Topology of the pathways to the inferior olive: an HRP study in cat. Neuroscience Letters 10: 207–214

Burne RA, Mihailoff GA, Woodward DJ (1978) Visual corticopontine input to the paraflocculus: a combined autoradiographic and horseradish peroxidase study. Brain Res 143: 139–146

Carpenter MB, Hanna GR (1961) Fiber projections from the spinal trigeminal nucleus in the cat. J Comp Neurol 117: 117–131

Cedarbaum JM, Aghajanian GK (1978) Afferent projections to the rat locus coeruleus as determined by a retrograde tracing technique. J Comp Neurol 178: 1–16

Chan-Palay V (1975) Fine structure of labelled axons in the cerebellar cortex and nuclei of rodents and primates after intraventricular infusions with tritiated serotonin. Anat Embryol (Berl) 148: 235–265

Chan-Palay V (1976) Serotonin afferents from raphe nuclei to the cerebellum. 7th International Conference on "Afferent and intrinsic organization of laminated structures in the brain." Exp Brain Res [Suppl] 1: 20–25

Chan-Palay V (1977) Cerebellar dentate nucleus. Organization, cytology and transmitters. Springer, Berlin-Heidelberg

Chan-Palay V (1978) The paratrigeminal nucleus. I. Neurons and synaptic organization. J Neurocytol 7: 405–418

Cheek MD, Rustioni A, Trevino DL (1975) Dorsal column nuclei projections to the cerebellar cortex in cats as revealed by the use of the retrograde transport of horseradish peroxidase. J Comp Neurol 164: 31–46

Chu NS, Bloom FE (1974) The catecholamine-containing neurons in the cat dorsolateral pontine tegmentum: distribution of the cell bodies and some axonal projections. Brain Res 66: 1–21

Cohen B, Goto K, Shanzer S, Weiss AH (1965) Eye movements induced by electric stimulation of the cerebellum in the alert cat. Exp Neurol 13: 145–162

Conrad LCA, Leonard CM, Pfaff DW (1974) Connections of the median and dorsal raphe nuclei in the rat: an autoradiographic and degeneration study. J Comp Neurol 156: 179–206

Cooke JD, Larson B, Oscarsson O, Sjölund B (1971) Origin and termination of cuneocerebellar tract. Exp Brain Res 13: 339–358

Corvaja N, Pompeiano O (1979) Identification of cerebellar corticovestibular neurons retrogradely labelled with horseradish peroxidase. Neuroscience 4:507–515

Corvaja N, Grofová I, Pompeiano O, Walberg F (1977) The lateral reticular nucleus in the cat – I. An experimental anatomical study of its spinal and supraspinal afferent connections. Neuroscience 2: 537–553

Courville J (1966) Rubrobulbar fibres to the facial nucleus and the lateral reticular nucleus (nucleus of the lateral funiculus). An experimental study in the cat with silver impregnation methods. Brain Res 1: 317–337

Courville J, (1975) Distribution of olivocerebellar fibers demonstrated by a radioautographic tracing method. Brain Res 95: 253–263

Courville J, Diakiw N (1976) Cerebellar corticonuclear projection in the cat. The vermis of the anterior and posterior lobes. Brain Res 110: 1–20

Courville J, Faraco-Cantin F (1976) Cerebellar corticonuclear projection demonstrated by the horseradish peroxidase method. Neuroscience Abstracts 2: 108

Courville J, Faraco-Cantin F (1978) On the origin of the climbing fibers of the cerebellum. An experimental study in the cat with an autoradiographic tracing method. Neuroscience 3: 797–809

Courville J, Faraco-Cantin F (to be published) Topography of the olivo-cerebellar projection. An experimental study in the cat with an autoradiographic tracing method.

Courville J, Diakiw N, Brodal A (1973) Cerebellar corticonuclear projection in the cat. The paramedian lobule. An experimental study with silver methods. Brain Res 50: 25–45

Cupedo RNJ (1965) A trigeminal midbrain-cerebellar fiber connection in the rat. J Comp Neurol 124: 61–70

Dahlström A, Fuxe K (1964) Evidence for the existence of monoamine-containing neurons in the central nervous system. I. Demonstration of monoamines in the cell bodies of brain stem neurons. Acta Physiol Scand [Suppl] 232: 1–80

Dekker JJ (1979) Ultrastructural characterization of cerebellar terminals on rubrospinal neurons in the rat. A combined electronmicroscopic study using anterograde and retrograde axoplasmic transport. Neuroscience Abstracts 5: 361

Desclin JC (1974) Histological evidence supporting the inferior olive as the major source of cerebellar climbing fibers in the rat. Brain Res 77: 365–384

Dietrichs E, Walberg F (1979) The cerebellar projection from the lateral reticular nucleus as studied with retrograde transport of horseradish peroxidase. Anat Embryol (Berl) 155: 273–290

Edwards SB (1972) The ascending and descending projections of the red nucleus in the cat: an experimental study using an autoradiographic tracing method. Brain Res 48: 45–63

Edwards SB, Rosenquist SC, Palmer LA (1974) An autoradiographic study of ventral lateral geniculate projections in the cat. Brain Res 72: 282–287

Eisenmann LM (1976) Horseradish peroxidase studies of brainstem projections to the vermis and paramedian lobule of the cerebellar cortex in the rat. Anat Rec 184: 396

Eller T, Chan-Palay V (1976) Afferents to the cerebellar lateral nucleus. Evidence from retrograde transport of horseradish peroxidase after pressure injections through micropipettes. J Comp Neurol 166: 285–302

Faull RLM (1977) A comparative study of the cells of origin of cerebellar afferents in the rat, cat, and monkey studied with the horseradish peroxidase technique. I. The non-vestibular brainstem afferents. Anat Rec 187: 577

Faull RLM (1978) The cerebellofugal projections in the brachium conjunctivum of the rat. II. The ipsilateral and contralateral descending pathways. J Comp Neurol 178: 519–536

Finger TE, Karten HJ (1977) The accessory optic system in teleost fishes. Neuroscience Abstracts 3: 90

Foote WE, Taber-Pierce E, Edwards L (1978) Evidence for a retinal projection to the midbrain raphe of the cat. Brain Res 156: 135–140

Gacek RR (1969) The course and central termination of first order neurons supplying vestibular endorgans in the cat. Acta Otolaryngol [Suppl] (Stockh) 254: 1–66

Gacek RR (1977) Location of brain stem neurons projecting to the oculomotor nucleus in the cat. Exp Neurol 57: 725–749

Gacek RR (1978) Location of commissural neurons in the vestibular nuclei of the cat. Exp Neurol 59: 479–491

Gacek RR (1979a) Location of abducens afferent neurons in the cat. Exp Neurol 64: 342–353

Gacek RR (1979b) Location of trochlear vestibuloocular neurons in the cat. Exp Neurol 66:692–706

Gallager DW, Pert A (1978) Afferents to brain stem nuclei (brain stem raphe, nucleus retivularis pontis caudalis and nucleus gigantocellularis) in the rat as demonstrated by microiontophoretically applied horseradish peroxidase. Brain Res 144: 257–275

Geisert EE Jr (1976) The use of tritiated horseradish peroxidase for defining neuronal pathways: a new application. Brain Res 117: 130–135

Glickstein M, Stein J, King RA (1972) Visual input to the pontine nuclei. Science 178: 1110–1111

Gordon G, Horrobin D (1967) Antidromic and synaptic responses in the cat's gracile nucleus to cerebellar stimulation. Brain Res 5: 419–421

Gould BB (1979) The organization of afferents to the cerebellar cortex in the cat: Projections from the deep cerebellar nuclei. J Comp Neurol 184: 27–42

Gould BB, Graybiel AM (1976) Afferents to the cerebellar cortex in the cat: evidence for an intrinsic pathway leading from the deep nuclei to the cortex. Brain Res 110: 601–611

Graham J (1977) An autoradiographic study of the efferent connections of the superior colliculus in the cat. J Comp Neurol 173: 629–654

Graham RC JR, Karnovsky MJ (1966) The early stages of absorption of injected horseradish peroxidase in the proximal tubules of mouse kidney: ultrastructural cytochemistry by a new technique. J Histochem Cytochem 14: 291–302

Grant G (1962) Projection of the external cuneate nucleus onto the cerebellum in the cat: an experimental study using silver methods. Exp Neurol 5: 179–195

Graybiel AM (1974) Visuo-cerebellar and cerebello-visual connections involving the ventral lateral geniculate nucleus. Exp Brain Res 20: 303–306

Graybiel AM (1977) Direct and indirect preoculomotor pathways of the brainstem: an autoradiographic study of the pontine reticular formation in the cat. J Comp Neurol 175: 37–78

Graybiel AM, Hartwieg EA (1974) Some afferent connections of the oculomotor complex in the cat: an experimental study with tracer techniques. Brain Res 81: 543–551

Graybiel AM, Nauta HJW, Lasek RJ, Nauta WJH (1973) A cerebello-olivary pathway in the cat: an experimental study using autoradiographic tracing techniques. Brain Res 58: 205–211

Groenewegen HJ, Voogd J (1977) The parasagittal zonation within the olivocerebellar projection. I. Climbing fiber distribution in the vermis of cat cerebellum. J Comp Neurol 174: 417–488

Groenewegen HJ, Boesten AJP, Voogd J (1975) The dorsal column nuclear projections to the nucleus ventralis posterior lateralis thalami and the inferior olive in the cat: an autoradiographic J Comp Neurol 162: 505–518

Groenewegen HJ, Voogd J, Freedman SL (1979) The parasagittal zonation within the olivocerebellar projection. II. Climbing fiber distribution in the intermediate and hemispheric parts of cat cerebellum. J Comp Neurol 183: 551–602

Haines DE (1977) Cerebellar corticonuclear and cortico-vestibular fibers of the flocculonodular lobe in a prosmian primate (Galago senegalensis). J Comp Neurol 174: 607–630

Halaris AE, Jones BE, Moore R (1976) Axonal transport in serotonin neurons of the midbrain raphe. Brain Res 107: 555–574

Hampson JL (1949) Relationships between cat cerebral and cerebellar cortices. J Neurophysiol 12: 37–50

Hayes NL, Rustioni A (1979) Dual projections of single neurons are visualized simultaneously: use enzymatically inactive [^{3}H] HRP. Brain Res 165: 321–326

Hoddevik GH (1975) The pontocerebellar projection onto the paramedian lobule in the cat: an experimental study with the use of horseradish peroxidase as a tracer. Brain Res 95: 291–307

Hoddevik GH (1977) The pontine projection to the flocculonodular lobe and the paraflocculus studied by means of retrograde axonal transport of horseradish peroxidase in the rabbit. Exp Brain Res 30: 511–526

Hoddevik GH (1978) The projection from nucleus reticularis tegmenti pontis onto the cerebellum in the cat. Anat Embryol (Berl) 153: 227–242

Hoddevik GH, Brodal A (1977) The olivocerebellar projection studied with the method of retrograde axonal transport of horseradish peroxidase. V. The projections to the flocculonodular lobe and the paraflocculus in the rabbit. J Comp Neurol 176: 269–280

Hoddevik GH, Walberg F (1979) The pontine projection onto longitudinal zones of the paramedian lobule in the cat. Exp Brain Res 34: 233–240

Hoddevik GH, Brodal A, Walberg F (1976) The olivocerebellar projection in the cat studied with the method of retrograde axonal transport of horseradish peroxidase. III. The projection to the vermal visual area. J Comp Neurol 169: 155–170

Hoddevik GH, Brodal A, Kawamura K, Hashikawa T (1977) The pontine projection to the cerebellar vermal visual area studied by means of the retrograde axonal transport of horseradish peroxidase. Brain Res 123: 209–227

Hoffmann KP, Behrend K, Schoppmann A (1976) A direct afferent visual pathway from the nucleus of the optic tract to the inferior olive in the cat. Brain Res 115: 150–153

Hökfelt T, Fuxe K (1969) Cerebellar monoamine nerve terminals, a new type of afferent fibres to the cortex cerebelli. Exp Brain Res 9: 63–72

Ikeda M (1979) Projections from the spinal and the principal sensory nuclei of the trigeminal nerve to the cerebellar cortex in the cat, as studied by retrograde transport of horseradish peroxidase. J Comp Neurol 184: 567–586

Ito M (1972) Neural design of the cerebellar motor control system. Brain Res 40: 81–84

Jansen J, Brodal A (1940) Experimental studies on the intrinsic fibers of the cerebellum. II. The cortico-nuclear projection. J Comp Neurol 73: 267–321

Jones BE, Moore RY (1974) Catecholamine-containing neurons of the nucleus locus coeruleus in the cat. J Comp Neurol 157: 43–52

Karamanlidis A (1968) Trigemino-cerebellar fiber connections in the goat studied by means of the retrograde cell degeneration method. J Comp Neurol 133: 71–88

Kawamura K (1975) The pontine projection from the inferior colliculus in the cat. An experimental anatomical study. Brain Res 95: 309–322

Kawamura K, Brodal A (1973) The tectopontine projection in the cat: an experimental anatomical study with comments on pathways for teleceptive impulses to the cerebellum. J Comp Neurol 149: 371–390

Kawamura K, Chiba M (1979) Cortical neurons projecting to the pontine nuclei in the cat. An experimental study with the horseradish peroxidase technique. Exp Brain Res 35: 269–285

Kawamura K, Hashikawa T (1979) Olivocerebellar projections in the cat studied by means of anterograde axonal transport of labeled amino acids as tracers. Neuroscience 4: 1615–1633

Kawamura K, Brodal A, Hoddevik G (1974) The projection of the superior colliculus onto the reticular formation of the brain stem: an experimental anatomical study in the cat. Exp Brain Res 19: 1–19

Kimoto Y, Satoh K, Sakumoto T, Tohyama M, Shimizu N (1978) Afferent fiber connections from the lower brain stem to the rat cerebellum by the horseradish peroxidase method combined with MAO staining, with special reference to noradrenergic neurons. J Hirnforsch 19: 85–100

Kooy D van der, Kuypers HGJM, Catsman-Berrevoets CE (1978) Single mammillary body cells with divergent axon collaterals. Demonstration by a simple, fluorescent retrograde double labeling technique in the rat. Brain Res 158: 189–196

Korte GE (1979) The brainstem projection of the vestibular nerve in the cat. J Comp Neurol 184: 279–292

Korte GE, Mugnaini E (1979) The cerebellar projection of the vestibular nerve in the cat. J Comp Neurol 184: 265–278

Kotchabhakdi N, Walberg F (1977) Cerebellar afferents from neurons in motor nuclei of cranial nerves demonstrated by retrograde axonal transport of horseradish peroxidase. Brain Res 137: 158–163

Kotchabhakdi N, Walberg F (1978a) Primary vestibular afferent projections to the cerebellum as demonstrated by retrograde axonal transport of horseradish peroxidase. Brain Res 142: 142–146

Kotchabhakdi N, Walberg F (1978b) Cerebellar afferent projections from the vestibular nuclei in the cat: an experimental study with the method of retrograde axonal transport of horseradish peroxidase. Exp Brain Res 31: 591–604

Kotchabhakdi N, Hoddevik GH, Walberg F (1978) Cerebellar afferent projections from the perihypoglossal nuclei: an experimental study with the method of retrograde axonal transport of horseradish peroxidase. Exp Brain Res 31: 13–29

Kotchabhakdi N, Walberg F, Brodal A (1978) The olivocerebellar projection in the cat studied with the method of retrograde axonal transport of horseradish peroxidase. VII. The projection to lobulus simplex, crus I and II. J Comp Neurol 182: 293–314

Künzle H (1973) The topographic organization of spinal afferents to the lateral reticular nucleus of the cat. J Comp Neurol 149: 103–116

Künzle H (1975) Autoradiographic tracing of the cerebellar projections from the lateral reticular nucleus in the cat. Exp Brain Res 22: 255–266

Künzle H, Cuénod M (1973) Differential uptake of $[^3H]$ proline and $[^3H]$ leucine by neurons: its importance for the autoradiographic tracing of pathways. Brain Res 62: 213–217

Kuypers HGJM (1958) An anatomical analysis of cortico-bulbar connexions to the pons and lower brain stem in the cat. J Anat 92: 198–218

Ladpli R, Brodal A (1968) Experimental studies of commissural and reticular formation projections from the vestibular nuclei in the cat. Brain Res 8: 65–96

Larsell O (1947) The development of the cerebellum in man in relation to its comparative anatomy. J Comp Neurol 87: 85–127

Larsell O (1953) The cerebellum of the cat and the monkey. J Comp Neurol 99: 135–200

Larsell O (1970) The comparative anatomy and histology of the cerebellum from monotremes through apes. In: Jansen J (ed) University of Minnesota Press, Minneapolis, p 185

Larsell O, Jansen J (1972) The comparative anatomy and histology of the cerebellum: The human cerebellum, cerebellar connections, and cerebellar cortes. University of Minnesota Press, Minneapolis, pp 90–133

Linauts M, Martin GF (1978) The organization of olivo-cerebellar projections in the opossum, Didelphis virginiana, as revealed by the retrograde transport of horseradish peroxidase. J Comp Neurol 179: 355–382

Liu CN (1956) Afferent nerves to Clarke's and the lateral cuneate nuclei in the cat. Arch Neurol 75: 66–77

Maciewicz RJ, Eagen K, Kaneko CRS, Highstein SM (1977) Vestibular and medullary brain stem afferents to the abducens nucleus in the cat. Brain Res 123: 229–240

Maeda T, Pin C, Salvert D, Ligier M, Jouvet M (1973) Les neurones contenant des catécholamines du tegmentum pontique et leurs voies de projection chez le chat. Brain Res 57: 119–152

Maekawa K, Simpson JI (1973) Climbing fiber responses evoked in vestibulocerebellum from visual system. J Neurophysiol 36: 649–666

Maekawa K, Takeda T (1975) Mossy fiber responses evoked in the cerebellar flocculus of rabbits by stimulation of the optic pathway. Brain Res 98: 590–595

Marburg O (1945) Nucleus eminentiae teretis, corpus pontobulbare, and their fiber connections. Studies in abnormally developed and pathologic cases. J Neuropathol Exp Neurol 4: 195–216

Martin GF, Linauts M, Walker JM (1977) The nucleus corporis pontobulbaris of the North American opossum. J Comp Neurol 175: 345–372

Matsushita M, Ikeda M (1975) The central cervical nucleus as cell origin of a spinocerebellar tract arising from the cervical cord: a study in the cat using horseradish peroxidase. Brain Res 100: 412–417

Matsushita M, Ikeda M (1976) Projections from the lateral reticular nucleus to the cerebellar cortex and nuclei in the cat. Exp Brain Res 24: 403–421

Matsushita M, Hosoya Y, Ikeda M (1979) Anatomical organization of the spinocerebellar system in the cat, as studied by retrograde transport of horseradish peroxidase. J Comp Neurol 184: 81–106

Mehler WR (1969) Some neurological species differences – a posteriori. Ann NY Acad Sci 167: 424–468

Mehler WR (1977) A comparative study of the cells of origin of cerebellar afferents in the rat, cat and monkey studied with the horseradish peroxidase technique. II. The vestibular nuclear complex. Anat Rec 186:653

Mizuno N, Mochizuki K, Akimoto C, Matsushima R (1973) Pretectal projections to the inferior olive in the rabbit. Exp Neurol 39: 498–506

Mizuno N, Mochizuki K, Akimoto C, Matsushima R, Sasaki K (1973) Projections from the parietal cortex to the brain stem nuclei in the cat, with special reference to the parietal cerebro-cerebellar system. J Comp Neurol 147: 511–522

Morin F, Kennedy DT, Gardner E (1966) Spinal afferents to the lateral reticular nucleus. I. An histological study. J Comp Neurol 126: 511–522

76

Nauta HJW, Pritz MB, Lasek RJ (1974) Afferents to the rat caudoputamen studied with horsera-dish peroxidase, An evaluation of a retrograde neuroanatomical research method. Brain Res 67: 219–238

Nauta WJH (1958) Hippocampal projections and related neural pathways to the midbrain in the cat Brain 81– 319–341

Olson L, Fuxe K (1971) On the projections from the locus coeruleus noradrenaline neurons: the cerebellar innervation. Brain Res 28: 165–171

Pearson AA (1949) Further observations on the mesencephalic root of the trigeminal nerve. J Comp Neurol 91: 147–194

Phelan K, Mehler WR (1979) Vestibulo-cerebellar afferents from the inferior olive and the vestibu-lar nuclear complex. Neuroscience Abstracts 5: 106

Pickel VM, Krebs H, Bloom FE (1973) Proliferation of norepinephrine-containing axons in rat cerebellar cortex after peduncle lesions. Brain Res 59: 169–179

Pickel VM, Segal M, Bloom FE (1974) A radioautographic study of the efferent pathways of the nucleus locus coeruleus. J Comp Neurol 155: 15–42

Pin C, Jones B, Jouvet M (1968) Topographie des neurones monoaminergiques du tronc cérébral du chat: étude par histofluorescence. CR Soc Biol (Paris) 162: 2136–2141

Poitras D, Parent A (1978) Atlas of the distribution of monoamine-containing nerve cell bodies in the brain stem of the cat. J Comp Neurol 179: 699–718

Pompeiano O, Brodal A (1957) Spino-vestibular fibers in the cat. An experimental study. J Comp Neurol 108: 353–382

Pompeiano O, Mergner T, Corvaja N (1978) Commissural, perihypoglossal and reticular afferent projections to the vestibular nuclei in the cat. An experimental anatomical study with hor-seradish peroxidase. Arch Ital Biol 116: 130–172

Precht W, Volkind R, Blanks RHI (1977) Functional organization of the vestibular input to the anterior and posterior cerebellar vermis of cat. Exp Brain Res 27: 143–160

Ramón y Cajal S (1903) La doble via descendente nacida del pedunculo cerebeloso superior. Trab Lab Invest Biol Univ Madrid 2: 23–29

Reiner A, Karten HJ (1978) A bisynaptic retinocerebellar pathway in the turtle. Brain Res 150: 163–169

Rinvik E, Walberg F (1975) Studies on the cerebellar projections from the main and external cuneate nuclei in the cat by means of retrograde axonal transport of horseradish peroxidase. Brain Res 95: 371–381

Ron S, Robinson DA (1973) Eye movements evoked by cerebellar stimulation in the alert monkey. J Neurophysiol 36: 1004–1022

Rosina A, Provini L, Vitali A, Bentivoglio M, Kuypers HGJM (1979) Divergence within the pontine projection to the neocerebellum as revealed by retrograde fluorescent double labelling tech-nique. Neuroscience Letters Suppl 3: 128

Rubertone JA, Haines DE (1979) Vestibular projections to the flocculonodular lobe of a prosimian primate (Galago senegalensis). Anat Rec 193: 670–671

Ruegg DG, Eldred E, Weisendanger M (1978) Spinal projection to the dorsolateral nucleus of the caudal basilar pons in the cat. J Comp Neurol 179: 383–392

Ruggiero D, Batton RR III, Jayaraman A, Carpenter MB (1977) Brain stem afferents to the fasti-gial nucleus in the cat demonstrated by transport of horseradish peroxidase. J Comp Neurol 172: 180–210

Rustioni A, Macchi G (1968) Distribution of dorsal root fibers in the medulla oblongata of the cat. J. Comp Neurol 134: 113–126

Saint-Cyr JA, Courville J (1979) Projections from the vestibular nuclei to the inferior olive in the cat. An autoradiographic and horseradish peroxidase study. Brain Res 165: 189–200

Sakai K, Salvert D, Touret M, Jouvet M (1977) Afferent connections of the nucleus raphe dorsalis in the cat as visualized by the horseradish peroxidase technique. Brain Res 137: 11–35

Sakai K, Touret M, Salvert D, Leger L, Jouvet M (1977) Afferent projections to the cat locus coeruleus as visualized by the horseradish peroxidase technique. Brain Res 119: 21–41

Sanides D, Fries W, Albus K (1978) The corticopontine projection from the visual cortex of the cat: an autoradiographic investigation. J Comp Neurol 179: 77–88

Shinnar S, Maciewicz RJ, Shofer RJ (1975) A raphe projection to cat cerebellar cortex. Brain Res 97: 139–143

Smith RL (1975) Axonal projections and connections of the principal sensory trigeminal nucleus in the monkey. J Comp Neurol 163: 347–376

Snider RS (1975) A cerebellar-ceruleus pathway. Brain Res 88: 59–63

Snider R, Eldred E (1951) Electro-anatomical studies on cerebro-cerebellar connections in the cat. J Comp Neurol 95: 1–16

Snider RS, Stowell A (1944) Receiving areas of the tactile, auditory, and visual systems in the cerebellum. J Neurophysiol 7: 331–357

Snyder RL, Faull RLM, Mehler WR (1978) A comparative study of the neurons of origin of the spinocerebellar afferents in rat, cat, and squirrel monkey based on the retrograde transport of horseradish peroxidase. J Comp Neurol 181: 833–852

Somana R, Walberg F (1978a) Cerebellar afferents from the paramedian reticular nucleus studied with retrograde transport of horseradish peroxidase. Anat Embryol (Berl) 154: 353–368

Somana R, Walberg F (1978b) The cerebellar projection from locus coeruleus as studied with retrograde transport of horseradish peroxidase in the cat. Anat Embryol (Berl) 155: 87–94

Somana R, Walberg F (1979a) Cerebellar afferents from the nucleus of the solitary tract. Neuroscience Letters 11: 41–47

Somana R, Walberg F (1979b) The cerebellar projection from the parabrachial nucleus in the cat. Brain Res 172: 144–149

Somana R, Walberg F (1979c) The cerebellar projection from the paratrigeminal nucleus in the cat. Neuroscience Letters 15: 49–54

Sousa-Pinto A (1970) The cortical projection onto the paramedian reticular and perihypoglossal nuclei of the medulla oblongata of the cat. An experimental anatomical study. Brain Res 18: 77–91

Souso-Pinto A, Brodal A (1969) Demonstration of a somatotopical pattern in the cortico-olivary projection in the cat. An experimental anatomical study. Exp Brain Res 8: 364–386

Steindler DA (1977) Trigemino-cerebellar projections in normal and reeler mutant mice. Neuroscience Letters 6: 293–300

Steward O, Scoville SA, Vinsant SL (1977) Analysis of collateral projections with a double retrograde labeling technique. Neuroscience Letters 5: 1–5

Stewart WA, King RB (1963) Fiber projections from the nucleus caudalis of the spinal trigeminal nucleus. J Comp Neurol 121: 271–286

Taber E (1961) The cytoarchitecture of the brain stem of the cat. I. Brain stem nuclei of the cat. J Comp Neurol 116: 27–69

Taber E, Brodal A, Walberg F (1960) The raphe nuclei of the brain stem in the cat. I. Normal topography and cytoarchitecture and general discussion. J Comp Neurol 114: 161–187

Taber Pierce E, Foote WE, Hobson JA (1976) The efferent connections of the nucleus raphe dorsalis. Brain Res 107: 137–144

Taber Pierce E, Hoddevik GH, Walberg F (1977) The cerebellar projection from the raphe nuclei in the cat as studied with the method of retrograde transport of horseradish peroxidase. Anat Embryol (Berl) 152: 73–87

Takeda T, Maekawa K (1976) The origin of the pretecto-olivary tract. A study using the horseradish peroxidase method. Brain Res 117: 319–325

Thomas DM, Kaufman RP, Sprague JM, Chambers WW (1956) Experimental studies of the vermal cerebellar projections in the brain stem of the cat (fastigiobulbar tract). J Anat 90: 371–385

Tohyama M (1976) Comparative anatomy of cerebellar catecholamine innervation from teleosts to mammals. J Hirnforsch 17: 43–60

Tolbert DL, Massopust LC, Murphy MG, Young PA (1976) The anatomical organization of the cerebello-olivary projection in the cat. J Comp Neurol 170: 525–544

Torvik A, Brodal A (1954) The cerebellar projection of the perihypoglossal nuclei (nucleus intercalatus, nucleus praepositus hypoglossi, and nucleus of Roller) in the cat. J Neuropathol Exp Neurol 13: 515–527

Usunoff KG, Hassler R, Wagner A, Bak IJ (1974) The efferent connections of the head of the caudate nucleus in the cat, an experimental morphological study with special reference to a projection to the raphe nuclei. Brain Res 74: 143–148

Voogd J (1964) The cerebellum of the cat: structure and fibre connexions. Van Gorcum's thesis, Assen

Voogd J (1969) The importance of fiber connections in the comparative anatomy of the mammalian cerebellum. In: Llinas R (ed) Neurobiology of cerebellar evolution and development. Am Medical Assn, Chicago, pp 493–515

V'alberg F (1958) Descending connections to the lateral reticular nucelus. An experimental study in the cat. J Comp Neurol 109: 363–389

Walberg F (1961) Fastigiofugal fibers to the perihypoglossal nuclei in the cat. Exp Neurol 3: 525–541

Walberg F (1974) Descending connections from the mesencephalon to the inferior olive: an experimental study in the cat. Exp Brain Res 21: 145–156

Walberg F, Brodal A (1979) The longitudinal zonal pattern in the paramedian lobule of the cat's cerebellum: an analysis based on a correlation of recent HRP data with results of studies with other methods. J Comp Neurol 187: 581–588

Walberg F, Jansen J (1964) Cerebellar corticonuclear projection studied experimentally with silver impregnation methods. J Hirnforsch 6: 338–345

Walberg F, Pompeiano O (1960) Fastigiofugal fibres to the lateral reticular nucleus; an experimental study in the cat. Exp Neurol 2: 40–53

Walberg F, Bowsher D, Brodal A (1958) The termination of primary vestibular fibers in the vestibular nuclei in the cat. An experimental study with silver methods. J Comp Neurol 110: 391–419

Walberg F, Pompeiano O, Brodal A, Jansen J (1962) The fastigiovestibular projection in the cat. An experimental study with silver impregnation methods. J Comp Neurol 118: 49–76

Walberg F, Pompeiano O, Westrum LE, Hauglie-Hanssen E (1962) Fastigioreticular fibers in the cat. An experimental study with silver methods. J Comp Neurol 119: 187–199

Walberg F, Kotchabhakdi N, Hoddevik GH (1979) The olivocerebellar projections to the flocculus and paraflocculus in the cat, compared to those in the rabbit. A study using horseradish peroxidase as a tracer. Brain Res 161: 389–398

Watson CRR, Switzer RC III (1978) Trigeminal projections to cerebellar tactile areas in the rat-origin mainly from n. interpolaris and n. principalis. Neuroscience Letters 10: 77–82

Weber JT, Partlow GD, Harting JK (1978) The projection of the superior colliculus upon the inferior olivary complex of the cat: an autoradiographic and horseradish peroxidase study. Brain Res 144: 369–377

Weisberg JA, Metz CB (1976) Simultaneous visualization of chromatolytic neurons and neurons retrogradely labelled with horseradish peroxidase. Neuroscience Letters 3: 167–171

Wiksten B (1975) The central cervical nucleus – a source of spinocerebellar fibres, demonstrated by retrograde transport of horseradish peroxidase. Neuroscience Letters 1: 81–84

Winfield JA, Hendrickson A, Kimm J (1978) Anatomical evidence that the medial terminal nucleus of the accessory optic tract in mammals provides a visual mossy fiber input to the flocculus. Brain Res 151: 175–182

Woodburne RT (1936) A phylogenetic consideration of the primary and secondary centers and connections of the trigeminal complex in a series of vertebrates. J Comp Neurol 65: 403–501

Yamamoto M (1979) Topographical representation in rabbit cerebellar flocculus for various afferent inputs from the brainstem investigated by means of retrograde axonal transport of horseradish peroxidase. Neuroscience Letters 12: 29–34

Subject Index

Other Reviews of Interest in this Series

Part 6: **Lüdicke, M.**: Internal Ear
Angioarchitectonic of Serpents.
21 figures. 41 pages. 1978.
ISBN 3-540-08836-9

Volume 55

Part 1: **Reutter, K.**: Taste Organ
in the Bullhead (Teleostei).
20 figures. 98 pages. 1978.
ISBN 3-540-08880-6

Part 2: **Dvořák, M.**: The
Differentation of Rat Ova
During Cleavage. 62 figures.
131 pages. 1978.
ISBN 3-540-08983-7

Part 3: **Wagner, H.-J.**: Cell Types and
Connectivity Patterns in Mosaic Retinas.
30 figures. 81 pages. 1978.
ISBN 3-540-09013-4

Part 4: **Jones, D.G.**: Some Current
Concepts of Synaptic Organization.
21 figures. 69 pages. 1978.
ISBN 3-540-09011-8

Part 5: **Fleischer, G.**: Evolutionary
Principles of the Mammalian Middle
Ear. 25 figures. 70 pages. 1978.
ISBN 3-540-09140-8

Volume 56

Kaissling, B.; Kriz, W.:
Structural Analysis of the Rabbit Kidney.
47 figures. VIII, 123 pages. 1979.
ISBN 3-540-09145-9

Volume 57

Niimi, K., Matsuoka, H.:
Thalamocortical Organization of the
Auditory System in the Cat Studied by
Retrograde Axonal Transport of Horse-
radish Peroxidase. 30 figures.
X, 56 pages. 1979.
ISBN 3-540-09449-0

Volume 58

Verwoerd, C.D.A. van Oostrom, C.G.:
Cephalic Neural Crest and Placodes.
41 figures. VI, 75 pages. 1979.
ISBN 3-540-09608-6

Volume 59

Bär, T.: The Vascular System of the
Cerebral Cortex.
33 figures. VI, 60 pages. 1980.
ISBN 3-540-09652-3

Volume 60

Hildebrand, R.: Nuclear Volume
and Cellular Metabolism.
12 figures. VII, 54 pages. 1980.
ISBN 3-540-09796-1

Volume 61

Korr, H.: Proliferation of Different Cell
Types in the Brain.
21 figures. VII, 72 pages. 1980.
ISBN 3-540-09899-2

Springer-Verlag Berlin Heidelberg New York